Extensions of Linear–Quadratic Control, Optimization and Matrix Theory

This is Volume 133 in
MATHEMATICS IN SCIENCE AND ENGINEERING
A Series of Monographs and Textbooks
Edited by RICHARD BELLMAN, *University of Southern California*

The complete listing of books in this series is available from the Publisher upon request.

Extensions of Linear–Quadratic Control, Optimization and Matrix Theory

DAVID H. JACOBSON

National Research Institute for Mathematical Sciences
Council for Scientific and Industrial Research, South Africa
(Honorary Professor in the University of the Witwatersrand)

1977

Academic Press
London · New York · San Francisco
A Subsidiary of Harcourt Brace Jovanovich, Publishers

ACADEMIC PRESS INC. (LONDON) LTD
24/28 Oval Road
London NW1

United States Edition published by
ACADEMIC PRESS INC.
111 Fifth Avenue
New York, New York 10003

Library of Congress Catalog Card Number: 77–76545
ISBN 0–12–378750–5

Printed in Great Britain by Galliard (Printers) Ltd, Great Yarmouth, Norfolk

PREFACE

Control, optimization and matrix theory are closely linked in many ways, perhaps most strongly by the linear-quadratic aspects they have in common. The present work seeks to extend, develop and strengthen this link by presenting a number of extensions of the well-known linear-quadratic theories. Consequently it should prove to be particularly useful to graduate students, teachers and researchers in science and engineering.

In a very definite sense this is a personal volume - it reflects my attempts over the past five years to understand and analyse non-linear systems and to contribute new developments. Inevitably some of the material presented has previously appeared in one or another form elsewhere in the literature but many results are being made known here for the first time.

Certain of the results presented in Chapters 2, 3, 5 and 6 were developed jointly with Drs. J.L. Speyer, W.M. Getz, M. Pachter and C.A. Botsaris; their contribution and cooperation is gratefully acknowledged. Dr. D.J. Bell kindly perused the draft manuscript.

Warm thanks are due to the Executive of the Council for Scientific and Industrial Research and to Professor R.E. Bellman who encouraged the project, to Mr. F.R. Baudert and Dr. D. de Jongh who edited the draft manuscript and proofread the typescript, and to my secretary Elsa de Beer who expertly typed the volume.

David H. Jacobson
Pretoria, 1977

for

LYNNE

LISA ' SVEN

CONTENTS

1. INTRODUCTION

The treatments of linear-quadratic control problems given in [1] are probably the most comprehensive available, though by now a little dated in some respects. Both discrete-time and continuous-time formulations are treated in that reference, and variational and dynamic programming techniques are used in the generation of solutions. A perusal of [1] thus provides a rather good sample of formulations and techniques, in addition to a good list of textbooks and other references.

In this short chapter we do not very exhaustively review the material in [1] (most of which is also available elsewhere), but rather describe loosely but adequately what, for the purpose of this monograph, constitutes a linear-quadratic formulation in control, optimization or matrix theory. We then outline the contents of the following chapters in some detail in order to elucidate the magnitude of the 'extensions' presented.

Broadly speaking a linear-quadratic (Gaussian) control (variational) formulation consists of a finite-dimensional linear, discrete- or continuous-time, dynamic system which is to be controlled in such a way as to minimize the value (or expected value) of a performance criterion which is the integral, or sum, of quadratic functions of the system state and control variables plus, perhaps, a quadratic function of the state at the terminal time. Stochastic formulations allow additive Gaussian white noise to disturb the dynamic system, and the outputs that can be measured are assumed to be linear functions of the state and Gaussian white noise. The most celebrated property of the solution of the non-singular linear-quadratic-problem (often referred to as an LQP or LQG) is that the optimal control is a linear (time-varying) function of the state or, in the stochastic case, the best estimate of the state. The matrix Riccati equation and, in the singular case [2], matrix inequalities play a special role in ensuring the existence of the solution. Questions and assumptions relating to stability, controllability and observability of the system are also important here.

By a linear-quadratic formulation in matrix theory we mean the study of the properties of positive (semi-) definite quadratic functions of a finite number of variables and their relation to linear equalities and inequalities. Here it is simply the definiteness of the quadratic function that makes the formulation standard. If this assumption is relaxed we have immediately a non-convex quadratic function which has properties not shared by the convex (positive semi-definite) case.

We use the term 'optimization', as distinct from 'optimal control', to describe the problem of finding a minimum of a function of a finite number of variables subject to equality and inequality constraints. 'Linear-quadratic' in this context refers to the fact that algorithms for the solution of the minimization problem are almost always based upon a model based in turn on the assumptions that the function to be minimized is a positive-definite quadratic form and that the constraints are linear.

It is fairly evident from the foregoing descriptions that the linear-quadratic thread that runs through control, optimization and matrix theory forms a strong conceptual and operational tie between them. Consequently in this monograph control, optimization and matrix theory are not strictly confined to separate chapters: in fact each of them is concerned with these three subjects of study to a greater or lesser extent.

In Chapter 2 we 'extend' the linear-quadratic control problem by first replacing the quadratic performance criterion by the exponential of a quadratic function. In the deterministic case we gain nothing by this move, as minimization of an exponential of a function is equivalent to minimization of that functional, but in the stochastic case a new, interesting formulation results. If the state is perfectly measurable but Gaussian white noise enters linearly into the linear system, we find that the optimal feedback controller is linear, as in the linear-quadratic case, but that the controller depends upon the statistics of the noise, unlike that for the linear-quadratic case. It turns out, interestingly, that the controller is equivalent to that obtained when the noise is treated as a 'belligerent player' in a two-person zero-sum linear-quadratic game, and this provides new justification for this type of 'worst case' controller design. If the measurement of the state is noise-corrupted, the optimal feedback controller retains its linear character but is in the general case infinite-dimensional. This is another surprise when it is recalled that in the linear-quadratic Gaussian case the controller turns out to be the finite-dimensional optimal controller for the deterministic case simply with the state replaced by the best (Kalman) filtered estimate of the state. An application due to Speyer of the exponential formulation and solution to homing missile guidance is also mentioned.

The next 'extension' is obtained by generalizing the linear dynamic system to a class of restricted non-linear stochastic systems while retaining the quadratic performance criterion. The optimal controller is here linear in the system state but depends upon the noise parameters. Known results due to Wonham, McLane and Kleinmann for linear systems with multiplicative noise are generalized here.

Next, we turn to the class of non-linear systems homogeneous-in-the-input. We demonstrate that such systems are asymptotically stabilizable under certain conditions which are almost necessary and sufficient. Furthermore, we show that stabilizing controllers actually minimize a wide variety of non-quadratic performance criteria.

We also obtain the solution to the problem of minimizing a certain non-quadratic performance criterion subject to a linear dynamic constraint. This result is generalized by Speyer to a stochastic version which involves control of a linear stochastic dynamic system driven by additive and state-dependent white-noise processes.

Finally in this chapter we study the control of systems of quadratic and bilinear differential equations and obtain some limited results, viz. that for a certain class of problems the optimal feedback controller is linear.

Taken as a whole, Chapter 2 illustrates that the linear-quadratic control problem has been extended in non-trivial ways both by using performance criteria more general than quadratic and by introducing classes of non-linear dynamic systems. These give rise to both linear and non-linear controllers.

In Chapter 3 we begin with matrix theory. First, copositive matrices are introduced. Quite simply a symmetric matrix is copositive if its associated quadratic form is non-negative for all vectors having non-negative elements. Interestingly, it turns out that all copositive matrices are sums of positive semi-definite matrices and matrices with non-negative elements (non-negative matrices) if and only if the dimensionality of the matrix is less than five. We show that this also implies that all positive semi-definite non-negative matrices have non-negative factorizations if and only if they are of dimension less than five. We show further that the representation for copositive matrices extends beyond dimension five if a more general type of copositivity, viz. stochastic copositivity, is defined.

Closely related to copositive quadratic forms is the question of non-negativity of a quadratic form subject to equality and inequality quadratic constraints. In the case of one constraint Finsler's theorem provides a complete answer, and in the case of an arbitrary number of constraints we extend Finsler's theorem to provide a useful sufficient condition. We use this extension to yield insight into the properties of the inverse of copositive matrices.

We then turn to symmetric M-matrices which are in fact positive-definite, and whose inverses are both positive-definite and non-negative. We show that these inverses have non-negative factorizations.

Next we apply copositive matrix theory to the non-convex quadratic programming problem to provide sufficient conditions for optimality.

The remainder of Chapter 3 is concerned with a study of the behaviour of solutions of systems of autonomous quadratic differential equations. Specifically we develop two sets of sufficient conditions for a solution to exhibit a finite escape time. The first set is similar to certain conditions obtained by Freeman, while the second set, being based upon our results for non-convex quadratic programming derived earlier, is less restrictive owing to our non-trivial use of the notion of invariant sets.

Chapter 4 contains what we believe are significant extensions of our results in [2] for the non-negativity of quadratic functionals. First we review and reformulate certain important sufficient conditions for the non-negativity of unconstrained quadratic functionals and extend these to the case where the control variables are constrained. A novel Riccati differential equation results from this approach. Next we further extend these sufficient conditions to a general class of non-quadratic, non-linear, constrained problems. Our results bear a resemblance to certain controllability conditions derived by Kunzi and Davison, and allow us to relate the non-negativity of non-quadratic functionals to that of a class of non-autonomous quadratic functionals.

Chapter 5 is concerned with the controllability of autonomous linear dynamic systems in which the control variables are constrained to lie within a certain constraint set. It is well known that, provided zero belongs to the interior of the convex hull of the constraint set, such a linear system is null-controllable if and only if it is completely controllable when the constraint is removed. More recently Brammer has provided necessary and sufficient conditions for null-controllability when zero does not belong to the interior of the convex hull of the constraint set. Arbitrary-interval null-controllability, introduced in Chapter 5, requires that the system be controllable on any time interval, this being a more demanding requirement than null-controllability. As is well known, a system is arbitrary-interval null-controllable if it is null-controllable and if zero belongs to the interior of the convex hull of the constraint set. The main purpose of Chapter 5, then, is to provide necessary and sufficient conditions for arbitrary-interval null-controllability when the constraint set is of general type. Most interesting is the role of arbitrary-interval null-controllability as a necessary and sufficient condition for continuity of the minimum time function in time-optimal control of an autonomous linear dynamic system.

In Chapter 6 we proceed to function minimization. We discuss the properties of a homogeneous model in comparison with a quadratic model and refer to a convergent algorithm for use on general functions. We also refer to the recent work of Kowalik who has further improved the effectiveness of the algorithm by introducing a highly stable numerical method in place of the Householder updating used in the first versions of the homogeneous algorithms.

Next in Chapter 6 we introduce the differential descent approach presented in [3] and further developed extensively by Botsaris. In this approach curvilinear, as opposed to linear search paths are used, which are developed by approximating the trajectories of steepest descent in appropriate ways. Such methods have considerable advantages in that they do not fail when Newton's method does, and automatically behave as gradient methods when far from the minimum of the function to be minimized, and as Newton's method when in the neighbourhood of the minimum.

Chapter 7 briefly assesses the earlier ones and indicates areas for further research.

1.1 References

[1] IEEE Transactions on Automatic Control, vol. AC-19, December 1971, 'Special Issue on the Linear-Quadratic-Gaussian Problem'.

[2] BELL, D.J. & JACOBSON, D.H. Singular Optimal Control Problems. Academic Press, New York and London, 1975.

[3] BOTSARIS, C.A. Differential Descent Methods for Function Minimization. Ph.D. Thesis, University of the Witwatersrand, Johannesburg, South Africa, 1975.

2. NON-LINEAR-QUADRATIC CONTROL PROBLEMS

2.1 Exponential Performance Criterion - Perfect Measurements

We consider in this section the optimal control of a linear discrete-time dynamic system disturbed by additive Gaussian white noise. In place of a quadratic performance criterion we use an exponential one [1]-[3]. We assume that the state of the system can be measured perfectly.

The assumption of Gaussian noise is deliberate - indeed it is the exponential nature of the Gaussian density function which matches the exponential nature of the performance criterion and results in the linear form of the optimal feedback controller.

2.1.1 Discrete-time Formulation

We consider a linear discrete-time dynamic system described by

$$x_{k+1} = A_k x_k + B_k u_k + \Gamma_k \omega_k, \quad k=0,\ldots,N-1; \quad x_0 \text{ given} \tag{2.1.1}$$

where the 'state' vector $x_k \in R^n$, the control vector $u_k \in R^m$, and the Gaussian noise input $\omega_k \in R^q$. The known matrices A_k, B_k, Γ_k have appropriate dimensions and may vary as a function of the index k.

The noise input is a sequence of independently distributed Gaussian random variables having probability density

$$p_\omega(\omega_0,\ldots,\omega_{N-1}) = \prod_{k=0}^{N-1} p(\omega_k;k) \tag{2.1.2}$$

where

$$p(\omega_k;k) = \frac{1}{\sqrt{(2\pi)^q|P_k^{-1}|}} \exp\{-\tfrac{1}{2}\omega_k^T P_k \omega_k\} \tag{2.1.3}$$

with

$$P_k > 0 \text{ (positive definite)}, \quad k=0,\ldots,N-1. \tag{2.1.4}$$

Note that

$$E[\omega_k] = 0, \quad E[\omega_k\omega_k^T] = P_k^{-1}, \quad k=0,\ldots,N-1 \tag{2.1.5}$$

where E denotes the expected value operator.

The performance criterion which we minimize in order to obtain a desirable controller for (2.1.1), is specified by

$$V^\sigma(x_o) = E_{|x_o}\,\sigma \exp\{\frac{\sigma}{2}[\sum_{k=o}^{N-1}(x_k^TQ_kx_k+u_k^TR_ku_k)+x_N^TQ_Nx_N]\} \tag{2.1.6}$$

where

$$Q_k \geqslant 0 \text{ (positive semi-definite)}, \quad k=0,\ldots,N \tag{2.1.7}$$

$$R_k > 0 \text{ (positive definite)}, \quad k=0,\ldots,N-1 \tag{2.1.8}$$

and where $\sigma = -1$ or $+1$.

We shall refer to (2.1.6) as a negative (positive) exponential performance criterion when $\sigma = -1(+1)$.

We wish to determine a Borel-measurable function C_k^σ such that the control policy

$$u_k^\sigma \triangleq C_k^\sigma(X_k), \quad k=0,\ldots,N-1; \quad X_k \triangleq \{x_o,x_1,\ldots,x_k\} \tag{2.1.9}$$

minimizes (2.1.6). Note that for an arbitrary choice of controls $\{u_k\}$, $V^-(x_o)$ and $V^+(x_o)$ are bounded according to

$$-1 \leqslant V^-(x_o) \leqslant 0, \quad 1 \leqslant V^+(x_o). \tag{2.1.10}$$

The inequalities (2.1.10) imply that $V^+(x_o)$ may become unbounded from above while $V^-(x_o)$ cannot.

Note that if there is no noise present, i.e.

$$\omega_k \equiv 0, \quad k=0,\ldots,N-1 \tag{2.1.11}$$

minimization of (2.1.6) is equivalent to minimization of

$$\sum_{k=o}^{N-1} (x_k^T Q_k x_k + u_k^T R_k u_k) + x_N^T Q_N x_N \tag{2.1.12}$$

subject to

$$x_{k+1} = A_k x_k + B_k u_k, \quad k=0,\ldots,N-1 \tag{2.1.13}$$

which is the standard linear-quadratic problem (LQP).

2.1.2 Discrete-time Solution

We define

$$J^\sigma(X_k;k) \triangleq \min_{u_k,\ldots,u_{N-1}} E_{|X_k} \sigma \exp\{\frac{\sigma}{2}[\sum_{i=k}^{N-1}(x_i^T Q_i x_i + u_i^T R_i u_i) + x_N^T Q_N x_N]\} \tag{2.1.14}$$

so that at time k+1

$$J^{\sigma}(X_{k+1};k+1) =$$

$$\min_{u_{k+1},\ldots,u_{N-1}} E_{|X_{k+1}} \sigma \exp\{\frac{\sigma}{2}[\sum_{i=k+1}^{N-1} (x_i^T Q_i x_i + u_i^T R_i u_i) + x_N^T Q_N x_N]\}. \tag{2.1.15}$$

Equations (2.1.14) and (2.1.15) yield

$$J^{\sigma}(X_k;k) = \min_{u_k} [\exp\{\frac{\sigma}{2}(x_k^T Q_k x_k + u_k^T R_k u_k)\} E_{|X_k} J^{\sigma}(X_{k+1};k+1)] \tag{2.1.16}$$

where

$$x_{k+1} = A_k x_k + B_k u_k + \Gamma_k \omega_k, \quad x_k \text{ given.} \tag{2.1.17}$$

The conditional expectation in (2.1.16) may now be written explicitly to yield

$$J^{\sigma}(X_k;k) = \min_{u_k} [\exp\{\frac{\sigma}{2}(x_k^T Q_k x_k + u_k^T R_k u_k)\} \int_{-\infty}^{\infty} p(\omega_k;k) J^{\sigma}(X_{k+1};k+1) d\omega_k] \tag{2.1.18}$$

and the boundary condition at k=N follows from (2.1.14) as

$$J^{\sigma}(X_N;N) = \sigma \exp\{\frac{\sigma}{2} x_N^T Q_N x_N\}. \tag{2.1.19}$$

We now prove that

$$J^{\sigma}(X_k;k) \triangleq \sigma F_k^{\sigma} \exp\{\frac{\sigma}{2} x_k^T W_k^{\sigma} x_k\}, \tag{2.1.20}$$

which is defined for $k=0,\ldots,N$, solves (2.1.18) where $W_k^{\sigma} \geqslant 0$, $k=0,\ldots,N$, is given by

$$W_k^{\sigma} = Q_k + A_k^T [\tilde{W}_{k+1}^{\sigma} - \tilde{W}_{k+1}^{\sigma} B_k (R_k + B_k^T \tilde{W}_{k+1}^{\sigma} B_k)^{-1} B_k^T \tilde{W}_{k+1}^{\sigma}] A_k \tag{2.1.21}$$

where

$$\tilde{W}^{\sigma}_{k+1} = W^{\sigma}_{k+1} + \sigma W^{\sigma}_{k+1}\Gamma_k(P_k-\sigma\Gamma^T_k W^{\sigma}_{k+1}\Gamma_k)^{-1}\Gamma^T_k W^{\sigma}_{k+1} \tag{2.1.22}$$

and where

$$W^{\sigma}_N = Q_N. \tag{2.1.23}$$

In addition, we have that

$$F^{\sigma}_k = F^{\sigma}_{k+1}\sqrt{\frac{|(P_k-\sigma\Gamma^T_k W^{\sigma}_{k+1}\Gamma_k)^{-1}|}{|P^{-1}_k|}}\ , \quad F^{\sigma}_N = 1 \tag{2.1.24}$$

and the optimal control policy is

$$u^{\sigma}_k = -C^{\sigma}_k x_k \tag{2.1.25}$$

where

$$C^{\sigma}_k = (R_k+B^T_k\tilde{W}^{\sigma}_{k+1}B_k)^{-1}B^T_k\tilde{W}^{\sigma}_{k+1}A_k, \quad k=0,\ldots,N-1. \tag{2.1.26}$$

We need the following Lemma in the proof that (2.1.20) solves (2.1.18). Note that the Lemma depends critically upon the exponential nature of $p(\omega_k;k)$.

LEMMA If $(P_k-\sigma\Gamma^T_k W^{\sigma}_{k+1}\Gamma_k) > 0$, then

$$\int_{-\infty}^{\infty} \frac{1}{\sqrt{(2\pi)^q|P^{-1}_k|}}\exp\{-\tfrac{1}{2}\omega^T_k P_k\omega_k\}.\exp\{\tfrac{\sigma}{2}\,x^T_{k+1}W^{\sigma}_{k+1}x_{k+1}\}d\omega_k$$

$$= \sqrt{\frac{|(P_k-\sigma\Gamma^T_k W^{\sigma}_{k+1}\Gamma_k)^{-1}|}{|P^{-1}_k|}}\ .\exp\{\tfrac{\sigma}{2}(A_k x_k+B_k u_k)^T\tilde{W}^{\sigma}_{k+1}(A_k x_k+B_k u_k)\}. \tag{2.1.27}$$

PROOF OF LEMMA The left-hand side of (2.1.27) is equal to

$$\int_{-\infty}^{\infty} \frac{1}{\sqrt{(2\pi)^q |P_k^{-1}|}}$$

$$.\exp\{\tfrac{1}{2}[(A_k x_k + B_k u_k + \Gamma_k \omega_k)^T \sigma W_{k+1}^{\sigma} (A_k x_k + B_k u_k + \Gamma_k \omega_k) - \omega_k^T P_k \omega_k]\} d\omega_k \tag{2.1.28}$$

which is equal to

$$\sqrt{\frac{|(P_k - \sigma\Gamma_k^T W_{k+1}^{\sigma} \Gamma_k)^{-1}|}{|P_k^{-1}|}} .\exp\{\tfrac{\sigma}{2}(A_k x_k + B_k u_k)^T \tilde{W}_{k+1}^{\sigma} (A_k x_k + B_k u_k)\}$$

$$. \int_{-\infty}^{\infty} \frac{1}{\sqrt{(2\pi)^q |(P_k - \sigma\Gamma_k^T W_{k+1}^{\sigma} \Gamma_k)^{-1}|}}$$

$$.\exp\{-\tfrac{1}{2}(\omega_k - \bar{\omega}_k)^T (P_k - \sigma\Gamma_k^T W_{k+1}^{\sigma} \Gamma_k)(\omega_k - \bar{\omega}_k)\} d\omega_k \tag{2.1.29}$$

where

$$\bar{\omega}_k \triangleq \sigma(P_k - \sigma\Gamma_k^T W_{k+1}^{\sigma} \Gamma_k)^{-1} \Gamma_k^T W_{k+1}^{\sigma} (A_k x_k + B_k u_k). \tag{2.1.30}$$

The Lemma is proved since the integrand in (2.1.29) is a probability density function.

Returning now to equation (2.1.18), and using (2.1.20) and the Lemma, we see that its right-hand side becomes

$$F_k^{\sigma}.\min_{u_k}[\sigma \exp\{\frac{\sigma}{2}(x_k^T Q_k x_k + u_k^T R_k u_k)\}$$

$$.\exp\{\frac{\sigma}{2}(A_k x_k + B_k u_k)^T \tilde{W}_{k+1}^{\sigma}(A_k x_k + B_k u_k)\}] \tag{2.1.31}$$

which upon minimization and simplification yields

$$\sigma F_k^{\sigma} \exp\{\frac{\sigma}{2} x_k^T W_k^{\sigma} x_k\} \tag{2.1.32}$$

which proves, by induction, that

$$J^{\sigma}(X_k;k) = \sigma F_k^{\sigma} \exp\{\frac{\sigma}{2} x_k^T W_k^{\sigma} x_k\}, \quad k=0,\dots,N. \tag{2.1.33}$$

It is easy to verify by induction that under assumptions (2.1.4), (2.1.7) and (2.1.8) W_k^- and $\tilde{W}_k^-$ are well-defined and positive semi-definite and that

$$(R_k + B_k^T \tilde{W}_{k+1}^- B_k) > 0, \quad k=0,\dots,N-1. \tag{2.1.34}$$

The case when $\sigma = +1$ is discussed below.

2.1.3 Properties of Discrete-time Solution ($\sigma = -1$)

The optimal feedback solution

$$u_k^- = -C_k^- x_k, \quad k=0,\dots,N-1 \tag{2.1.35}$$

is a linear function of the system state. This controller, unlike that for the linear-quadratic problem, depends upon P_k^{-1}, the covariance matrix of the additive system noise.

It is interesting to investigate here two limiting cases: the first in which $\lambda_{min}(P_k) \to \infty$ ('no noise'); and the second in which $\lambda_{min}(P_k^{-1}) \to \infty$ ('infinitely wild noise'). In the first case it is easy to verify that $C_k^- \to D_k$, the optimal feedback gain matrix for the deterministic linear-quadratic problem. In the second case we shall assume that $\Gamma_k^T Q_{k+1} \Gamma_k > 0$, k=0,...,N-1, so that $\Gamma_k^T W_{k+1}^- \Gamma_k > 0$, k=0,...,N-1. As $P_k^{-1} \to 0$ we then have that

$$\tilde{W}_{k+1}^- \to W_{k+1}^- - W_{k+1}^- \Gamma_k^T (\Gamma_k^T W_{k+1}^- \Gamma_k)^{-1} \Gamma_k^T W_{k+1}^-, \quad k=0,\ldots,N-1 \tag{2.1.36}$$

and, from (2.1.20), (2.1.24) it follows that

$$J^-(x_0;0) \to 0. \tag{2.1.37}$$

Note also that if Γ_k has rank n for k=0,...,N-1,

$$\tilde{W}_{k+1}^- \to 0, \quad k=0,\ldots,N-1 \tag{2.1.38}$$

and

$$C_k^- \to 0, \quad k=0,\ldots,N-1. \tag{2.1.39}$$

In words, if all components of x_k are disturbed by 'infinitely wild additive noise' there is no point (as far as performance criterion $V^-(x_0)$ is concerned) in exercising control to try to counteract these 'infinite unpredictable disturbances'.

2.1.4 Properties of Discrete-time Solution ($\sigma = +1$)

Here the (linear) optimal feedback controller is

$$u_k^+ = -C_k^+ x_k, \quad k=0,\ldots,N-1 \tag{2.1.40}$$

and again we note that the controller is directly affected by the noise covariance P_k^{-1} via C_k^+. When $\lambda_{min}(P_k) \to \infty$ ('no noise') we see that $C_k^+ \to D_k$, as is the case when $\sigma = -1$. However, if $\lambda_{min}(P_k^{-1})$ becomes large, $J^+(x_o;0)$ can cease to exist. To see this, we assume that

$$\Gamma_k^T Q_{k+1} \Gamma_k > 0, \; k=0,\ldots,N-1 \tag{2.1.41}$$

and that

$$P_j - \Gamma_j^T W_{j+1}^+ \Gamma_j > 0, \quad j=k+1,\ldots,N-1. \tag{2.1.42}$$

It then follows that

$$\Gamma_k^T W_{k+1}^+ \Gamma_k > 0 \tag{2.1.43}$$

so that for $\lambda_{min}(P_k^{-1})$ sufficiently large

$$P_k - \Gamma_k^T W_{k+1}^+ \Gamma_k \not> 0, \tag{2.1.44}$$

which clearly implies that the left-hand side of (2.1.27) is 'infinite'. Clearly, then $J^+(x_k;k)$ is 'infinite', as are $J^+(x_o;0)$ and $V^+(x_o)$.

2.1.5 Relation to Discrete-time Games

When $\sigma = -1$ the solution (2.1.25) is easily seen to be identical to the solution of the following cooperative deterministic game (actually, a linear-quadratic control problem):

$$\underset{\{u_k\},\{\omega_k\}}{\text{Minimize}} \; [\tfrac{1}{2} \sum_{k=0}^{N-1} (x_k^T Q_k x_k + u_k^T R_k u_k + \omega_k^T P_k \omega_k) + \tfrac{1}{2} x_N^T Q_N x_N] \quad (2.1.45)$$

subject to

$$x_{k+1} = A_k x_k + B_k u_k + \Gamma_k \omega_k \, , \quad k=0,\dots,N-1, \quad x_0 \text{ given.} \quad (2.1.46)$$

Note that the above formulation determines a linear optimal feedback strategy-pair,

$$u_k^- = -C_k^- x_k, \quad \omega_k^- = -\Delta_k^- x_k, \quad k=0,\dots,N-1. \quad (2.1.47)$$

We now have a new interpretation for the linear-quadratic cooperative game. If player u_k <u>assumes</u> that player ω_k will cooperate in minimizing the quadratic criterion, the optimal feedback controller (policy) that is obtained for u_k, namely $u_k^- = -C_k^- x_k$, is also optimal relative to an exponential performance criterion, if ω_k is Gaussian white noise.

When $\sigma = +1$, the appropriate game is non-cooperative, viz.

$$\underset{\{u_k\}}{\text{Min}} \; \underset{\{\omega_k\}}{\text{Max}} \; [\tfrac{1}{2} \sum_{k=0}^{N-1} (x_k^T Q_k x_k + u_k^T R_k u_k - \omega_k^T P_k \omega_k) + \tfrac{1}{2} x_N^T Q_N x_N] \quad (2.1.48)$$

subject to (2.1.46).

It is well known, and easy to verify, that if

$$P_k - \Gamma_k^T W_{k+1}^+ \Gamma_k > 0, \quad k=0,\ldots,N-1 \qquad (2.1.49)$$

the value of (2.1.48) is given by $\frac{1}{2}x_0^T W_0^+ x_0$. However, if $P_k - \Gamma_k^T W_{k+1}^+ \Gamma_k$ is indefinite, (2.1.48) does not have a finite value.

Our interpretation of the above non-cooperative deterministic game is as follows. If player u_k assumes that ω_k will be uncooperative in minimizing the quadratic criterion, the optimal feedback strategy for u_k, namely $u_k^+ = -C_k^+ x_k$, is also optimal relative to an exponential performance criterion, if ω_k is Gaussian white noise. When looked at from this viewpoint the min-max game solution for u_k('worst-case design') does not appear to be too pessimistic, since the exponential performance criterion is rather appealing.

2.1.6 Continuous-time Formulation

Here the performance criterion is

$$V^\sigma(x_0) = E_{|x_0}\, \sigma \exp\{\frac{\sigma}{2}[\int_0^T (x^TQx+u^TRu)dt+x^T(T)Q_Tx(T)]\} \qquad (2.1.50)$$

subject to the linear stochastic differential equation

$$dx = Axdt + Budt + \Gamma d\alpha \qquad (2.1.51)$$

where, for notational simplicity, the time-dependence of the variables has been suppressed. Note that $Q \geqslant 0$, $R > 0$, for all $t \in [0,T]$, and that $Q_T \geqslant 0$.

In (2.1.51) $\alpha(\cdot)$ is a Brownian motion process of zero mean and satisfies

$$E[\alpha(t)\alpha^T(\tau)] = \int_0^{\min(t,\tau)} P^{-1}(s)ds \qquad (2.1.52)$$

where $P > 0$ for all $t \in [0,T]$.

Formally, (2.1.51) may be considered as

$$\dot{x} = Ax + Bu + \Gamma\omega \qquad (2.1.53)$$

where ω, the formal derivative of α, is zero mean Gaussian white noise and satisfies

$$E[\omega(t)\omega^T(\tau)] = P^{-1}.\delta(t-\tau) \qquad (2.1.54)$$

and where δ is the 'Dirac delta function'.

We seek a Borel-measurable function C^σ such that

$$u^\sigma(X,t) = C^\sigma(X,t), \quad t \in [0,T] \qquad (2.1.55)$$

minimizes (2.1.50), where

$$X \triangleq \{x(\tau) | 0 \leqslant \tau \leqslant t\}. \qquad (2.1.56)$$

2.1.7 Continuous-time Solution

The continuous-time problem may be solved either by formally taking the limit of the solutions of the discrete-time versions or by solving the 'generalized' Hamilton-Jacobi-Bellman equation for the optimal value function $J^\sigma(x,t)$, viz.

$$-\frac{\partial J^{\sigma}(x;t)}{\partial t} = \min_{u} \{\frac{\sigma}{2}(x^TQx+u^TRu)J^{\sigma}(x,t)+[J_x^{\sigma}(x,t)]^T(Ax+Bu)$$

$$+ \tfrac{1}{2}\,tr[J_{xx}^{\sigma}(x,t)\Gamma P^{-1}\Gamma^T]\}. \qquad (2.1.57)$$

Using either method we find that

$$u^{\sigma}(x,t) = -R^{-1}B^TS^{\sigma}x, \quad t \in [0,T] \qquad (2.1.58)$$

and

$$J^{\sigma}(x,t) = \sigma F^{\sigma} \exp\{\frac{\sigma}{2}\, x^TS^{\sigma}x\} \qquad (2.1.59)$$

where

$$-\dot{S}^{\sigma} = Q + S^{\sigma}A + A^TS^{\sigma} - S^{\sigma}(BR^{-1}B^T-\sigma\Gamma P^{-1}\Gamma^T)S^{\sigma}, \quad S^{\sigma}(T) = Q_T \qquad (2.1.60)$$

and

$$-\dot{F}^{\sigma} = \tfrac{1}{2}\sigma F^{\sigma}tr(S^{\sigma}\Gamma P^{-1}\Gamma^T), \quad F^{\sigma}(T) = 1. \qquad (2.1.61)$$

2.1.8 Relation to Differential Games

It is not hard to verify that (2.1.58) with $\sigma = -1$ is also the optimal policy for controller u in the following differential game,

$$\underset{u(\cdot),\omega(\cdot)}{\text{Minimize}} \; [\tfrac{1}{2}\int_0^T (x^TQx+u^TRu+\omega^TP\omega)dt+\tfrac{1}{2}x^T(T)Q_Tx(T)] \qquad (2.1.62)$$

subject to

$$\dot{x} = Ax + Bu + \Gamma\omega, \quad x_0 \text{ given}. \qquad (2.1.63)$$

Note that because of our assumptions on Q, R, P, Q_T the

solution $S^-(\cdot)$ of the Riccati equation (2.1.60) exists for all $t \in [0,T]$.

When $\sigma = +1$ the appropriate game is non-cooperative, viz.

$$\underset{u(\cdot)}{\text{Min}}\ \underset{\omega(\cdot)}{\text{Max}}\ [\tfrac{1}{2}\int_0^T (x^TQx+u^TRu-\omega^TP\omega)dt+\tfrac{1}{2}x^T(T)Q_Tx(T)] \tag{2.1.64}$$

subject to (2.1.63). In this case the value of (2.1.64) is equal to $\frac{1}{2}x^T(0)S^+(0)x(0)$ provided that the following Riccati equation has a solution on $[0,T]$:

$$-\dot{S}^+ = Q + S^+A + A^TS^+ - S^+(BR^{-1}B^T-\Gamma P^{-1}\Gamma^T)S^+, \quad S^+(T) = Q_T. \tag{2.1.65}$$

Note that by standard results on Riccati differential equations, (2.1.65) has a solution for all $t \in [0,T]$ if

$$(BR^{-1}B^T-\Gamma P^{-1}\Gamma^T) \geqslant 0, \quad \text{for all } t \in [0,T]. \tag{2.1.66}$$

If (2.1.66) is not satisfied (say for $\lambda_{min}(P^{-1})$ sufficiently large) the solution of equation (2.1.65) may exhibit a finite escape time, which would imply non-existence of a finite value of (2.1.50) and of (2.1.64).

2.1.9 Stability Properties

In this section we assume that all parameters are time-invariant and we investigate the stability of the system

$$\dot{x} = (A-BC_\infty^\sigma)x \tag{2.1.67}$$

where C_∞^σ is defined below.

First we consider the case when $\sigma = -1$. We assume that the pair (A,B) is completely controllable and that $Q > 0$. Then, standard theory guarantees the existence of S_∞^-, the unique positive-definite symmetric solution of the algebraic Riccati equation

$$Q + SA + A^TS - S(BR^{-1}B^T + \Gamma P^{-1}\Gamma^T)S = 0 \qquad (2.1.68)$$

and we have the time-invariant feedback gain

$$C_\infty^- = R^{-1}B^TS_\infty^- . \qquad (2.1.69)$$

We now define

$$L^- \triangleq \tfrac{1}{2}x^TS_\infty^-x \qquad (2.1.70)$$

which is positive definite. Along solutions of (2.1.67) we have

$$\dot{L}^- = \tfrac{1}{2}x^T(S_\infty^-A + A^TS_\infty^-)x - x^TS_\infty^-BR^{-1}B^TS_\infty^-x \qquad (2.1.71)$$

which, upon using (2.1.68), is

$$\dot{L}^- = -\tfrac{1}{2}x^T[Q + S_\infty^-(BR^{-1}B^T - \Gamma P^{-1}\Gamma^T)S_\infty^-]x. \qquad (2.1.72)$$

Now, if

$$BR^{-1}B^T - \Gamma P^{-1}\Gamma^T \geqslant 0 \qquad (2.1.73)$$

we have

$$\dot{L}^- < 0, \quad \text{for all } x \neq 0 \qquad (2.1.74)$$

so that L^- is a Liapunov function and system (2.1.67) with controller C_∞^- is asymptotically stable in the large. Note that simple examples demonstrate that (2.1.67) can be unstable if condition (2.1.73) is violated.

Next, we consider the case $\sigma = +1$. Here, we assume that condition (2.1.73) holds and also that $Q > 0$. Note that we can always define a matrix N by the formula

$$NN^T \triangleq BR^{-1}B^T - \Gamma P^{-1}\Gamma^T. \tag{2.1.75}$$

If we assume now that the pair (A,N) is completely controllable it follows that there exists a unique positive-definite symmetric matrix S_∞^+ satisfying

$$Q + SA + A^TS - S(BR^{-1}B^T - \Gamma P^{-1}\Gamma^T)S = 0 \tag{2.1.76}$$

and

$$C_\infty^+ = R^{-1}B^TS_\infty^+. \tag{2.1.77}$$

If we now define $L^+ \triangleq \frac{1}{2}x^TS_\infty^+x$, we can easily verify that L^+ is a Liapunov function and (2.1.67) with controller C_∞^+ is asymptotically stable in the large. Note the interesting point that condition (2.1.73) is sufficient, modulo complete controllability and definiteness of Q, to guarantee asymptotic stability of (2.1.67) with controllers C_∞^- or C_∞^+. In the first case it is used to guarantee negativity of $\dot{L}^-$, while in the second it guarantees the existence of S_∞^+.

2.2 Exponential Performance Criterion - Noisy Measurements

In Section 2.1.2 the controller that minimizes an exponential performance criterion was derived from a simple backward recursion (dynamic program). If, however, the system 'state' is not observed exactly, the multiplicative nature of the exponential performance criterion precludes a similar approach. Interestingly, it turns out that the optimal control is a linear functional of the smoothed history of the system 'state'. The feedback gains depend upon the statistics of both the system and the measurement noise.

2.2.1 Discrete-Time Formulation

The linear discrete-time dynamic system is specified by

$$x_{k+1} = A_k x_k + B_k u_k + \Gamma_k \omega_k, \quad k=0,\dots,N-1 \qquad (2.2.1)$$

where the 'state' vector $x_k \in R^n$, the control vector $u_k \in R^m$ and the Gaussian white-noise input $\omega_k \in R^q$. The known matrices A_k, B_k, Γ_k have appropriate dimensions and may vary as a function of the index k.

The Gaussian white 'process noise' ω_k has statistics

$$E[\omega_k] = 0, \quad E[\omega_j \omega_k^T] = W_k \delta_{jk}, \quad W_k \geqslant 0 \qquad (2.2.2)$$

and the initial 'state' x_0 is a Gaussian random variable with statistics

$$E[x_0] = 0, \quad E[x_0 x_0^T] = Y_0. \qquad (2.2.3)$$

At each stage, k, a noisy linear measurement z_k is made,

$$z_k = H_k x_k + \nu_k, \quad k=0,\dots,N \qquad (2.2.4)$$

where $z_k \in R^s$ and the random variable ν_k has statistics

$$E[\nu_k] = 0, \quad E[\nu_j \nu_k^T] = V_k \delta_{jk}, \quad V_k > 0. \tag{2.2.5}$$

The noise sequences $\{\omega_k\}$, $\{\nu_k\}$ and the initial 'state' x_0 are assumed to be independent of each other.

The performance criterion which we minimize in order to obtain a desirable controller for (2.2.1) is specified by

$$V^\sigma = E\sigma \exp\{\frac{\sigma}{2} \psi\} \tag{2.2.6}$$

where

$$\psi = \sum_{k=0}^{N-1} (x_k^T Q_k x_k + u_k^T R_k u_k) + x_N^T Q_N x_N \tag{2.2.7}$$

and where $Q_k \geqslant 0$, $k=0,\ldots,N$, $R_k > 0$, $k=0,\ldots,N-1$.

In (2.2.6), E denotes the expected value operator over the space of all the random variables (i.e. the 'unconditional expected value operator').

The problem, then, is to find a non-anticipative controller (i.e. one that is a Borel-measurable function of only the past measurements) that minimizes V^σ.

Defining

$$Z_k^T \triangleq [z_o^T, z_1^T, \ldots, z_k^T] \tag{2.2.8}$$

the control u_k may be any Borel-measurable function of Z_k. It follows that (2.2.6) can be written in terms of nested conditional expectations as follows:

$$V^{\sigma} = E[E_{|Z_o}[E_{|Z_1}[\ldots E_{|Z_N}[\sigma e^{\sigma\psi/2}]\ldots]]]. \qquad (2.2.9)$$

If we define

$$V^{\sigma}(Z_k) \triangleq E_{|Z_k}[\sigma e^{\sigma\psi/2}] \qquad (2.2.10)$$

where it is assumed that an admissible control sequence has been used, a recursion formula for $V^{\sigma}(Z_k)$ is obtained from (2.2.9) and (2.2.10) as

$$V^{\sigma}(Z_k) = E_{|Z_k}V^{\sigma}(Z_{k+1}), \quad k=0,\ldots,N-1. \qquad (2.2.11)$$

2.2.2 Discrete-time Solution - Terminal Cost Only

It turns out that it is easier to solve a terminal cost problem ($Q_k = 0$, $k=0,\ldots,N-1$) than the problem formulated above. We therefore first solve this simpler problem and then show how to convert the general problem into this form.

We have that

$$V^{\sigma}(Z_N) = E_{|Z_N}[\sigma \exp\{\tfrac{\sigma}{2}[\sum_{k=0}^{N-1} u_k^T R_k u_k + x_N^T Q_N x_N]\}] \qquad (2.2.12)$$

which may be written as follows, because the control u_k is a Borel-measurable function of the measurement history Z_k, $k=0,\ldots,N-1$,

$$V^{\sigma}(Z_N) = \sigma \exp\{\tfrac{\sigma}{2} \sum_{k=0}^{N-1} u_k^T R_k u_k\}.E_{|Z_N}[\exp\{\tfrac{\sigma}{2} x_N^T Q_N x_N\}]. \qquad (2.2.13)$$

In order to perform the expectation in (2.2.13) we require

the conditional probability density $p(x_N|Z_N)$. Now it is well known, see for example [4], that because of the linearity of (2.2.1) and (2.2.4) the conditional density $p(x_k|Z_k)$ is Gaussian with conditional mean $\hat{x}_k \triangleq E_{|Z_k}[x_k]$ and is propagated by the Kalman filter equations

$$\hat{x}_k = \bar{x}_k + K_k s_k \tag{2.2.14}$$

where

$$\bar{x}_k = E_{|Z_{k-1}}[x_k] = A_{k-1}\hat{x}_{k-1} + B_{k-1}u_{k-1} \tag{2.2.15}$$

and

$$s_k = z_k - H_k\bar{x}_k \, . \tag{2.2.16}$$

The gain matrix K_k is given by

$$K_k = P_k H_k^T V_k^{-1} \tag{2.2.17}$$

where the covariance $P_k \triangleq E[(x_k-\hat{x}_k)(x_k-\hat{x}_k)^T]$ is propagated by

$$P_k = M_k - M_k H_k^T (H_k M_k H_k^T + V_k)^{-1} H_k M_k \tag{2.2.18}$$

and where $M_k \triangleq E[(x_k-\bar{x}_k)(x_k-\bar{x}_k)^T]$ is propagated by

$$M_k = A_{k-1}P_{k-1}A_{k-1}^T + \Gamma_{k-1}W_{k-1}\Gamma_{k-1}^T . \tag{2.2.19}$$

The error in the estimate $e_k \triangleq x_k-\hat{x}_k$ is statistically independent of the measurement history Z_k and the estimate $\hat{x}_k$. The measurement residual s_k is Gaussian and is independent of the measurement history Z_{k-1}. We therefore conclude that $p(x_N|Z_N)$ is Gaussian with mean $\hat{x}_N$ and covariance P_N. Fortunately, the exponential form of $p(x_N|Z_N)$ allows us to compute the expectation in (2.2.13) in closed form (cf. the Lemma in Section 2.1.2), so that

$$V^{\sigma}(Z_N) = \sigma\alpha_N \exp\{\frac{\sigma}{2}[\sum_{k=0}^{N-1} u_k^T R_k u_k + \hat{x}_N^T \tilde{Q}_N \hat{x}_N]\} \qquad (2.2.20)$$

where

$$\tilde{Q}_N = Q_N + \sigma Q_N(P_N^{-1}-\sigma Q_N)^{-1}Q_N = Q_N[I-\sigma P_N Q_N]^{-1} \qquad (2.2.21)$$

and where we have assumed that $P_N^{-1}-\sigma Q_N > 0$. The coefficient α_N is given by

$$\alpha_N = [|(P_N^{-1}-\sigma Q_N)^{-1}|/|P_N|]^{\frac{1}{2}} = |I-\sigma P_N Q_N|^{-\frac{1}{2}}. \qquad (2.2.22)$$

A backward step is now taken using (2.2.11), namely

$$V^{\sigma}(Z_{N-1}) = E_{|Z_{N-1}} V^{\sigma}(Z_N). \qquad (2.2.23)$$

As u_k is a Borel-measurable function of Z_k, (2.2.23) can be written as

$$V^{\sigma}(Z_{N-1}) = \sigma\alpha_N \exp\{\frac{\sigma}{2} \sum_{k=0}^{N-1} u_k^T R_k u_k\}.E_{|Z_{N-1}} [\exp\{\frac{\sigma}{2} \hat{x}_N^T \tilde{Q}_N \hat{x}_N\}]. \qquad (2.2.24)$$

In (2.2.24) the conditional expectation is taken with respect to the measurement residual s_k at k=N where, at any stage k, s_k is Gaussian-distributed with zero mean and covariance

$$S_k \triangleq E[s_k s_k^T] = H_k M_k H_k^T + V_k. \tag{2.2.25}$$

Calculating the expectation in (2.2.24) we obtain

$$V^\sigma(Z_{N-1}) = \sigma\alpha_{N-1} \exp\{\tfrac{\sigma}{2}[\sum_{k=0}^{N-1} u_k^T R_k u_k + \bar{x}_N^T \bar{Q}_N \bar{x}_N]\} \tag{2.2.26}$$

where

$$\bar{Q}_N = \tilde{Q}_N + \sigma\tilde{Q}_N K_N (S_N^{-1} - \sigma K_N^T \tilde{Q}_N K_N)^{-1} K_N^T \tilde{Q}_N \tag{2.2.27}$$

and where we assume that $S_N^{-1} - \sigma K_N^T \tilde{Q}_N K_N > 0$. The expression for α_{N-1} is

$$\alpha_{N-1} = \alpha_N |I - \sigma K_N^T \tilde{Q}_N K_N|^{-\frac{1}{2}}. \tag{2.2.28}$$

Finally, using (2.2.15) in (2.2.26) we obtain

$$V^\sigma(Z_{N-1}) = \sigma\alpha_N \exp\{\tfrac{\sigma}{2}[\sum_{k=0}^{N-1} u_k^T R_k u_k$$

$$+ (A_{N-1}\hat{x}_{N-1} + B_{N-1}u_{N-1})^T \bar{Q}_N (A_{N-1}\hat{x}_{N-1} + B_{N-1}u_{N-1})]\}. \tag{2.2.29}$$

Now using (2.2.9) and the fundamental theorem in [4, p. 769] we have that the minimum value of V^σ, namely J^σ, is given by

$$J^\sigma \triangleq \min_{u_o,\ldots,u_{N-1}} E[V^\sigma(Z_{N-1})] = \min_{u_o,\ldots,u_{N-2}} E[\min_{u_{N-1}} V^\sigma(Z_{n-1})]. \tag{2.2.30}$$

Now,

$$J^\sigma(Z_{N-1}) \triangleq \min_{u_{N-1}} V^\sigma(Z_{N-1}) \tag{2.2.31}$$

is obtained by minimizing the right-hand side of (2.2.29) with respect to u_{N-1}, which yields

$$u_{N-1} = -(R_{N-1}+B_{N-1}^T\bar{Q}_N B_{N-1})^{-1}B_{N-1}^T\bar{Q}_N A_{N-1}\hat{x}_{N-1} \quad (2.2.32)$$

provided that $R_{N-1} + B_{N-1}^T\bar{Q}_N B_{N-1} > 0$.

It then follows that

$$J^{\sigma}(Z_{N-1}) = \sigma\alpha_{N-1}\ \exp\{\frac{\sigma}{2}[\sum_{k=0}^{N-2} u_k^T R_k u_k+\hat{x}_{N-1}^T\tilde{Q}_{N-1}\hat{x}_{N-1}]\} \quad (2.2.33)$$

where

$$\tilde{Q}_{N-1} = A_{N-1}^T[\bar{Q}_N-\bar{Q}_N B_{N-1}(B_{N-1}^T\bar{Q}_N B_{N-1}+R_{N-1})^{-1}B_{N-1}^T\bar{Q}_N]A_N. \quad (2.2.34)$$

Applying this procedure recursively backward in time yields at each stage the linear optimal feedback controller

$$u_k = -[R_k+B_k^T\bar{Q}_{k+1}B_k]^{-1}B_k^T\bar{Q}_{k+1}A_k\hat{x}_k, \quad k=0,\ldots,N-1 \quad (2.2.35)$$

on the assumption that $R_k + B_k^T\bar{Q}_{k+1}B_k > 0$, where

$$\bar{Q}_k = \tilde{Q}_k + \sigma\tilde{Q}_k K_k(S_k^{-1}-\sigma K_k^T\tilde{Q}_k K_k)^{-1}K_k^T\tilde{Q}_k \quad (2.2.36)$$

and where

$$\tilde{Q}_{k-1} = A_{k-1}^T[\bar{Q}_k-\bar{Q}_k B_{k-1}(B_{k-1}^T\bar{Q}_k B_{k-1}+R_{k-1})^{-1}B_{k-1}^T\bar{Q}_k]A_{k-1} \quad (2.2.37)$$

on the assumption that $S_k^{-1} - \sigma K_k^T\tilde{Q}_k K_k > 0$. The boundary condition for (2.2.36) and (2.2.37) is provided by (2.2.21).

The unconditional minimal expectation of the performance criterion V^{σ} is

$$J^{\sigma} = \sigma\alpha_0 \exp\{\frac{\sigma}{2} \bar{x}_0^T \bar{Q}_o \bar{x}_0\} \tag{2.2.38}$$

where α_0 is obtained via the backward recursion

$$\alpha_k = \alpha_{k+1} |I - \sigma K_{k+1} \tilde{Q}_{k+1} K_{k+1} S_{k+1}|^{-\frac{1}{2}} \tag{2.2.39}$$

with boundary condition provided by (2.2.22).

2.2.3 Discrete-time Solution - The General Case

When Q_k, k=0,...,N-1, are not all zero we see that (2.2.13) becomes

$$V^{\sigma}(Z_N) = \sigma \exp\{\frac{\sigma}{2} \sum_{k=0}^{N-1} u_k^T R_k u_k\} E_{|Z_N} [\exp\{\frac{\sigma}{2} \sum_{k=0}^{N} x_k^T Q_k x_k\}] \tag{2.2.40}$$

where the terms in u_k are legitimately taken ahead of the conditional expectation. Note that the expectation in (2.2.40), being conditional upon the whole measurement history, involves the smoothed estimate of x_k, k=0,...,N.

In order to convert this problem into the one solved in the previous section we first define

$$X_k^T \triangleq [x_0^T, x_1^T, \ldots, x_k^T] \tag{2.2.41}$$

where X_k is an n(k+1)-dimensional vector. Next, we define the (N+1)n x (N+1)n composite cost weighting matrix

$$L_N = \begin{pmatrix} Q_o & & & 0 \\ & Q_1 & & \\ & & \ddots & \\ 0 & & & Q_N \end{pmatrix} . \qquad (2.2.42)$$

Using (2.2.41) and (2.2.42), equation (2.2.40) becomes

$$V^{\sigma}(Z_N) = \sigma \exp\{\frac{\sigma}{2} \sum_{k=0}^{N-1} u_k^T R_k u_k\} E_{|Z_N} [\exp\{\frac{\sigma}{2} X_N^T L_N X_N\}] . \qquad (2.2.43)$$

Clearly, $p(X_N|Z_N)$ is required here in order to calculate the expectation in (2.2.43). First we see that the recurrence relation for X_k is

$$X_{k+1} = \tilde{A}_k X_k + \tilde{B}_k u_k + \tilde{\Gamma}_k \omega_k \qquad (2.2.44)$$

where

$$\tilde{A}_k = \begin{pmatrix} I_{(k+1)n} \\ 0 \quad A_k \end{pmatrix} , \quad \tilde{B}_k = \begin{pmatrix} 0 \\ B_k \end{pmatrix} , \quad \tilde{\Gamma}_k = \begin{pmatrix} 0 \\ \Gamma_k \end{pmatrix} \qquad (2.2.45)$$

and where $\tilde{A}_k \in R^{(k+2)n \times (k+1)n}$, $\tilde{B}_k \in R^{(k+2)n \times m}$, $\tilde{\Gamma}_k \in R^{(k+2)n \times q}$.

The measurement equation is

$$z_k = \tilde{H}_k X_k + \nu \qquad (2.2.46)$$

where

$$\tilde{H}_k^T = \begin{pmatrix} 0 \\ H_k^T \end{pmatrix} \qquad (2.2.47)$$

and where $\tilde{H}_k^T \varepsilon R^{(k+2)n \times s}$. Note that the dimensionality of X_k increases by n at each stage.

It may be verified that the Kalman filter theory is applicable to (2.2.44) and (2.2.46) and that $p(X_k|Z_k)$ is Gaussian with conditional mean $\hat{X}_k \triangleq E_{|Z_k}[X_k]$ and covariance $\tilde{P}_k \triangleq E_{|Z_k}[(X_k-\hat{X}_k)(X_k-\hat{X}_k)^T]$. The gain matrix of the Kalman filter is then

$$\tilde{K}_k \triangleq \tilde{P}_k\tilde{H}_k^T V_k^{-1} \tag{2.2.48}$$

and the Kalman filter is identical to (2.2.14)-(2.2.19) with x_k, $\bar{x}_k$, $\hat{x}_k$, K_k, A_k, B_k, P_k, M_k replaced by X_k, $\bar{X}_k$, $\hat{X}_k$, $\tilde{K}_k$, $\tilde{A}_k$, $\tilde{B}_k$, $\tilde{P}_k$, $\tilde{M}_k$.

We may now apply the results of Section 2.2.2 direct to obtain the linear optimal feedback controller as

$$u_k = -(R_k+\tilde{B}_k^T\bar{L}_{k+1}\tilde{B}_k)^{-1}\tilde{B}_k^T\bar{L}_{k+1}\tilde{A}_k\hat{X}_k, \quad k=0,\dots,N-1 \tag{2.2.49}$$

where

$$\bar{L}_{k+1} = \tilde{L}_{k+1} + \sigma\tilde{L}_{k+1}\tilde{K}_{k+1}(S_{k+1}^{-1}-\sigma\tilde{K}_{k+1}^T\tilde{L}_{k+1}\tilde{K}_{k+1})^{-1}\tilde{K}_{k+1}^T\tilde{L}_{k+1} \tag{2.2.50}$$

$$\tilde{L}_k = \tilde{A}_k^T[\bar{L}_{k+1}-\bar{L}_{k+1}\tilde{B}_k(\tilde{B}_k^T\bar{L}_{k+1}\tilde{B}_k+R_k)^{-1}\tilde{B}_k\bar{L}_{k+1}]\tilde{A}_k \tag{2.2.51}$$

$$\tilde{L}_N = L_N[I-\sigma\tilde{P}_N L_N]^{-1} \tag{2.2.52}$$

and where it is assumed that $(R_k+\tilde{B}_k^T\bar{L}_{k+1}\tilde{B}_k) > 0$, $(S_{k+1}^{-1}-\sigma\tilde{K}_{k+1}^T\tilde{L}_{k+1}\tilde{K}_{k+1}) > 0$ and $P_N^{-1}-\sigma L_N > 0$. Thus the optimal control is a linear function of the smoothed estimate of the entire state history.

Note that the unconditioned expected value of the performance criterion is

$$J^{\sigma} = \sigma\alpha_{\sigma} \exp\{\frac{\sigma}{2} \bar{x}_o^T \bar{L}_o \bar{x}_o\} \tag{2.2.53}$$

where α_o may be computed from

$$\alpha_k = \alpha_{k+1} | I - \sigma\tilde{K}_{k+1}^T \tilde{L}_{k+1} \tilde{K}_{k+1} S_{k+1} |^{-\frac{1}{2}} \tag{2.2.54}$$

and

$$\alpha_N = | I - \sigma L_N \tilde{P}_N |^{-\frac{1}{2}}. \tag{2.2.55}$$

2.2.4 No Process Noise

A simplification results if there is no process noise (i.e. $\omega_k = 0$, $k=0,\dots,N-1$) present. Indeed the problem can be transformed quite simply into a terminal cost problem with a state vector of dimension 2n. The optimal control at each stage is then a linear function of the 2n-dimensional state estimate.

Owing to the absence of process noise, we may write

$$x_k = \Phi_{N,k}^{-1}(x_N - \sum_{j=k}^{N-1} \Phi_{N,j+1} B_j u_j) \tag{2.2.56}$$

where the transition matrix $\Phi_{k,\ell}$ is defined as

$$\Phi_{k,\ell} = \prod_{j=\ell}^{k-1} A_j, \quad \Phi_{k,k} = I. \tag{2.2.57}$$

The quantity ψ then becomes

$$\psi = x_N^T C_N x_N - 2x_N^T [\sum_{k=0}^{N-1} C_k' \sum_{j=k}^{N-1} \Phi_{N,j} B_j u_j]$$

$$+ \sum_{k=0}^{N-1} [(\sum_{j=k}^{N-1} \Phi_{N,j+1} B_j u_j)^T C_k' (\sum_{j=k}^{N-1} \Phi_{N,j+1} B_j u_j)]$$

$$+ \sum_{k=0}^{N-1} u_k^T R_k u_k \tag{2.2.58}$$

where

$$C_k' = (\Phi_{N,k}^{-1})^T Q_k \Phi_{N,k}^{-1}, \quad C_k = \sum_{j=0}^{k} C_j'. \tag{2.2.59}$$

Equation (2.2.58) may be simplified to yield V^σ in the form

$$V^\sigma = \sigma E \exp\{\frac{\sigma}{2}[(x_N^T \; \eta_N^T) \begin{pmatrix} C_N & -I \\ -I & 0 \end{pmatrix} (x_N^T \; \eta_N^T)^T$$

$$+ \sum_{k=0}^{N-1} (\eta_k^T \; u_k^T) \begin{pmatrix} 0 & D_k \\ D_k^T & R_k + D_k^T C_k D_k \end{pmatrix} (\eta_k^T \; u_k^T)^T]\} \tag{2.2.60}$$

where

$$\eta_{k+1} = \eta_k + C_k \Phi_{N,k} B_k u_k, \quad \eta_o = 0 \tag{2.2.61}$$

and

$$D_k = \Phi_{N,k+1} B_k. \tag{2.2.62}$$

As η_k is a Borel-measurable function of Z_{k-1} the terms involving η_k and u_k can be taken in front of the conditioned expectation $E_{|Z_N}$ to yield

$$V^{\sigma}(Z_N) = \sigma\alpha_N \exp\{\frac{\sigma}{2}[\sum_{k=0}^{N-1}(\hat{y}_k^T\ u_k^T)\begin{pmatrix} 0 & \tilde{\tilde{D}}_k \\ \tilde{\tilde{D}}_k^T & \bar{R}_k \end{pmatrix}(\hat{y}_k^T\ u_k^T)^T$$

$$+ \hat{y}_N^T\tilde{\tilde{C}}_N\hat{y}_N]\} \qquad (2.2.63)$$

where

$$\hat{y}_k^T \triangleq [\hat{x}_k^T\ \eta_k^T]\ , \quad \tilde{\tilde{D}}_k \triangleq [0\ D_k^T], \quad \bar{R}_k \triangleq R_k + D_k^T C_k D_k. \qquad (2.2.64)$$

The estimation equations for $\hat{y}_k$ are

$$\hat{y}_k = \bar{y}_k + \tilde{\tilde{K}}_k s_k, \quad \bar{y}_{k+1} = \tilde{\tilde{A}}_k\hat{y}_k + \tilde{\tilde{B}}_k u_k \qquad (2.2.65)$$

where

$$\tilde{\tilde{K}}_k \triangleq \begin{pmatrix} K_k \\ 0 \end{pmatrix}, \quad \tilde{\tilde{A}}_k \triangleq \begin{pmatrix} A_k & 0 \\ 0 & I \end{pmatrix}, \quad \tilde{\tilde{B}}_k = \begin{pmatrix} B_k \\ C_k D_k \end{pmatrix}. \qquad (2.2.66)$$

As in the previous sections we can recursively determine the linear optimal feedback controller as

$$u_k = -(\bar{R}_k + \tilde{\tilde{B}}_k^T\bar{C}_{k+1}\tilde{\tilde{B}}_k)^{-1}(\tilde{\tilde{D}}_k^T + \tilde{\tilde{B}}_k^T\bar{C}_{k+1}\tilde{\tilde{A}}_k)\hat{y}_k \qquad (2.2.67)$$

where

$$\bar{C}_{k+1} = \tilde{\tilde{C}}_{k+1} + \sigma\tilde{\tilde{C}}_{k+1}\tilde{\tilde{K}}_{k+1}(S_{k+1}^{-1} - \sigma\tilde{\tilde{K}}_{k+1}^T\tilde{\tilde{C}}_{k+1}\tilde{\tilde{K}}_{k+1})^{-1}\tilde{\tilde{K}}_{k+1}^T\tilde{\tilde{C}}_{k+1} \qquad (2.2.68)$$

$$\tilde{\tilde{C}}_k = \tilde{\tilde{A}}_k^T\bar{C}_{k+1}\tilde{\tilde{A}}_k - (\tilde{\tilde{A}}_k^T\bar{C}_{k+1}\tilde{\tilde{B}}_k + \tilde{\tilde{D}}_k)(\bar{R}_k + \tilde{\tilde{B}}_k^T\bar{C}_{k+1}\tilde{\tilde{B}}_k)^{-1}(\tilde{\tilde{D}}_k^T + \tilde{\tilde{B}}_k^T\bar{C}_{k+1}\tilde{\tilde{A}}_k) \qquad (2.2.69)$$

and where

$$\tilde{\tilde{C}}_N = \begin{pmatrix} C_N(I-\sigma P_N C_N)^{-1} & -(I-\sigma C_N P_N)^{-1} \\ -(I-\sigma P_N C_N)^{-1} & \sigma(P_N^{-1}-\sigma C_N)^{-1} \end{pmatrix} . \qquad (2.2.70)$$

In the above equations it is assumed that $P_N^{-1}-\sigma C_N > 0$, $S_k^{-1}-\sigma\tilde{\tilde{K}}_k^T\tilde{\tilde{C}}_k\tilde{\tilde{K}}_k > 0$ and that $\bar{R}_k + \tilde{\tilde{B}}_k^T\bar{C}_{k+1}\tilde{\tilde{B}}_k > 0$.

The unconditional expected value of the performance criterion is given by

$$J^\sigma = \sigma\alpha_0 \, \exp\{\tfrac{\sigma}{2}(\bar{x}_0^T,0)\bar{C}_o(\bar{x}_0^T,0)^T\} \qquad (2.2.71)$$

where α_0 may be computed from

$$\alpha_k = \alpha_{k+1}|I-\sigma\tilde{\tilde{K}}_{k+1}^T\tilde{\tilde{C}}_{k+1}\tilde{\tilde{K}}_{k+1}S_{k+1}|^{-\frac{1}{2}} \qquad (2.2.72)$$

$$\alpha_N = |I-\sigma C_N P_N|^{-\frac{1}{2}} . \qquad (2.2.73)$$

2.2.5 Continuous-time Formulation

In continuous-time the performance criterion is

$$V^\sigma = \sigma E \, \exp\{\tfrac{\sigma}{2}[\int_0^T (x^TQx+u^TRu)dt+x^T(T)Q_Tx(T)]\} \quad (2.2.74)$$

subject to the linear stochastic differential equation

$$dx = Axdt + Budt + \Gamma d\alpha \qquad (2.2.75)$$

and the measurement equation

$$dz = Hxdt + d\beta \qquad (2.2.76)$$

where α and β are Brownian motion processes with zero mean that satisfy

$$E[\alpha(t)\alpha^T(\tau)] = \int_0^{\min(t,\tau)} W(s)ds \tag{2.2.77}$$

and

$$E[\beta(t)\beta^T(\tau)] = \int_0^{\min(t,\tau)} V(s)ds. \tag{2.2.78}$$

We assume that the initial state x_0 is Gaussian-distributed and that x_0, $\alpha(\cdot)$ and $\beta(\cdot)$ are independent.

The results for the terminal cost problem (i.e. $Q(t) = 0$, $t \in [0,T]$) may be found by taking the formal limit of the discrete-time equations. It turns out that the optimal feedback controller is

$$u = -R^{-1}B^T\tilde{Q}\hat{x}, \quad t \in [0,T] \tag{2.2.79}$$

where

$$-\dot{\tilde{Q}} = A^T\tilde{Q} + \tilde{Q}A - \tilde{Q}(BR^{-1}B^T - \sigma KVK^T)\tilde{Q} \tag{2.2.80}$$

$$\tilde{Q}(T) = Q_T[I - \sigma P(T)Q_T]^{-1} \tag{2.2.81}$$

where $\hat{x}$, the best estimate of the state, is propagated by the continuous form of the Kalman estimator which also yields the Kalman gain K and P, the covariance of the error in the estimate.

Although in the general case an infinite-dimensional controller results, we have that provided $\alpha(\cdot)$ is zero (no process noise) the results of Section 2.2.4 become

$$u = -R^{-1}(\tilde{\tilde{D}}^T + \tilde{\tilde{B}}^T\tilde{\tilde{C}})\hat{y} \tag{2.2.82}$$

where

$$-\dot{\tilde{\tilde{C}}} = \tilde{\tilde{A}}^T\tilde{\tilde{C}} + \widetilde{\widetilde{CA}} - (\widetilde{\widetilde{CB}}+\tilde{\tilde{D}})R^{-1}(\tilde{\tilde{D}}^T+\tilde{\tilde{B}}^T\tilde{\tilde{C}})+\sigma\widetilde{\widetilde{CK}}^T V\widetilde{\widetilde{KC}} \qquad (2.2.83)$$

where $\tilde{\tilde{C}}(T)$ is given by (2.2.70) with C_{N+1} replaced by $Q_T + C(T)$ and where $\tilde{\tilde{A}}$, $\tilde{\tilde{B}}$, $\tilde{\tilde{K}}$, $\tilde{\tilde{D}}$ and C are defined as

$$\tilde{\tilde{A}} \triangleq \begin{pmatrix} A & 0 \\ 0 & 0 \end{pmatrix}, \quad \tilde{\tilde{B}}^T \triangleq [B^T \; B^T\Phi^T(T,t)C^T], \quad \tilde{\tilde{K}}^T \triangleq [K^T \; 0],$$

$$\tilde{\tilde{D}}^T \triangleq [0 \; B^T\Phi^T(T,t)] \qquad (2.2.84)$$

$$C(t) \triangleq \int_0^t [\Phi^{-1}(T,\tau)]^T Q(\tau)\Phi^{-1}(T,\tau)d\tau. \qquad (2.2.85)$$

The transition matrix $\Phi(t,T)$ satisfies

$$\Phi^{-1}(T,t) \equiv \Phi(t,T), \quad \dot{\Phi}(t,T) = -\Phi(t,T)A, \quad \Phi(T,T) = I. \qquad (2.2.86)$$

The 2n vector $\hat{y} = [\hat{x}^T \; \eta^T]$ is propagated in part by the continuous-time Kalman filter and in part by

$$\dot{\eta} = C(t)\Phi(T,t)B(t)u(t), \quad \eta(0) = 0. \qquad (2.2.87)$$

2.2.6 A Note on Separation

We note that there is a separation [5] between estimation and control in the problems treated in Sections 2.1 and 2.2, namely that the estimators (Kalman filters) can be constructed without having the optimal feedback controller. All that is required is the input sequence $\{u_k\}$. The separation is, however, one-way, as the optimal controllers depend upon the estimation-error covariance. In contrast, the separation is both-ways in the Linear-Quadratic-Gaussian control problem where the parameters of the optimal feedback controller are dependent upon only the deterministic system and criterion matrices A, B, Q and R.

2.2.7 A Further Separation Property

The solution for the terminal cost problem in continuous time exhibits a further type of separation which allows an adaptive controller. This has been exploited [6] in an application of the exponential-criterion theory to homing missile guidance.

In this and other applications the measurement covariance matrix V is not known a priori and must be estimated on-line. In order to determine $\tilde{Q}$ the Riccati (2.2.80) and the Kalman filter covariance matrix equations would have to be solved on-line after V is estimated. However, the on-line backward integration of $\tilde{Q}$ may be avoided in the following way. First note that as V is estimated, the Kalman filter covariance equation

$$\dot{P} = AP + PA^T + \Gamma W \Gamma^T - PH^T V^{-1} HP, \quad P(t_0) = P_0 \tag{2.2.88}$$

must be integrated to yield P as a function of t. Next, note that $U \triangleq \tilde{Q}^{-1}$ satisfies

$$\dot{U} = UA^T + AU - BR^{-1}B^T + \sigma PH^T V^{-1} HP \tag{2.2.89}$$

with boundary condition

$$U(T) = Q_T^{-1} - \sigma P(T). \tag{2.2.90}$$

Now, define a new matrix variable

$$M \triangleq U + \sigma P \tag{2.2.91}$$

so that we have

$$\dot{M} = AM + MA^T - BR^{-1}B^T + \sigma\Gamma W\Gamma^T \tag{2.2.92}$$

with boundary condition

$$M(T) = Q_T^{-1}. \tag{2.2.93}$$

Letting $S \triangleq M^{-1}$, we see that

$$-\dot{S} = SA + A^TS - S(BR^{-1}B^T - \sigma\Gamma W\Gamma^T)S, \quad S(T) = Q_T. \tag{2.2.94}$$

It follows that if the process noise covariance W is known, $S(\cdot)$ can be calculated off-line and stored (note that S is identical to S^σ in Section 2.1.7). Then $\tilde{Q}$ can be calculated on-line using the formula

$$\tilde{Q} = (S^{-1} - \sigma P)^{-1} = (I - \sigma SP)^{-1}S \tag{2.2.95}$$

which is obtained from (2.2.91).

Thus the effect of the measurement noise covariance V on the control gains $\tilde{Q}$ is only via P. Naturally the adaptive gain is optimal only if V is estimated perfectly, which is usually not the case; however, the above procedure at least suggests a rational approach.

The above adaptive scheme has been applied in [6] and has been shown to exhibit certain advantages over the standard linear-quadratic design.

2.3 Non-linear Stochastic Systems

We turn now to a different extension of the linear-quadratic theory. We retain the quadratic performance criterion but extend the class of dynamic systems by adding a certain type of non-linear stochastic term [7] (this formulation includes that of linear systems with multiplicative noise [8], [9]). It turns out also for this class of problems that the optimal feedback controller is linear and that its coefficients depend upon the statistics of the noise.

2.3.1 Discrete-time Formulation

We consider the problem of minimizing with respect to $\{u_k\}$ the performance criterion

$$E_{|x_0}\left[\sum_{k=0}^{N-1} \tfrac{1}{2}(x_k^T Q_k x_k + u_k^T R_k u_k) + \tfrac{1}{2} x_N^T Q_N x_N\right] \tag{2.3.1}$$

subject to the stochastic dynamic equation

$$x_{k+1} = A_k x_k + B_k u_k + m_k + f_k(x_k, u_k, \omega_k), \quad x_0 \text{ given} \tag{2.3.2}$$

where $x_k \in R^n$, $u_k \in R^m$, $\omega_k \in R^q$, $f_k : R^n \times R^m \times R^q \to R^n$ and Q_N, Q_k, R_k, A_k, B_k are matrices of appropriate dimension, and where $m_k \in R^n$.

The noise sequence $\{\omega_k\}$ is assumed to be independently distributed in time and not necessarily Gaussian. Our assumptions on the random vector function f_k are as follows.

ASSUMPTION 2.3.1 $\bar{f}_k(x_k,u_k) \triangleq E[f_k(x_k,u_k,\omega_k)]$ is zero for all $x_k \in R^n$, $u_k \in R^m$, $k=0,\dots,N-1$.

Note that the results derived in this section hold also if $\bar{f}_k$ is a linear function of x_k and u_k. However, there is no loss of generality in assuming that $\bar{f}_k = 0$ because appropriate choices of A_k, B_k and m_k will model any mean value of f_k which is linear in x_k and u_k.

ASSUMPTION 2.3.2 $F_k(x_k,u_k) \triangleq E[f_k(x_k,u_k,\omega_k)f_k^T(x_k,u_k,\omega_k)]$ exists and is a general quadratic function of x_k, u_k for $k=0,\dots,N-1$. That is, $F_k(x_k,u_k)$ has the representation

$$F_k(x_k,u_k) = P_k^0 + \sum_{i=1}^{n'} P_k^i(\tfrac{1}{2}x_k^TW_k^ix_k+u_k^TN_k^ix_k+\tfrac{1}{2}u_k^TM_k^iu_k+x_k^Tg_k^i+u_k^Th_k^i) \qquad (2.3.3)$$

where P_k^i, W_k^i, N_k^i, M_k^i are matrices of appropriate dimensions, $g_k^i \in R^n$, $h_k^i \in R^m$, and P_k^i, W_k^i, M_k^i are symmetric, $i=0,\dots,n'$, where $n' = n(n+1)/2$.

ASSUMPTION 2.3.3 Representation (2.3.3) is such that $F_k(x_k,u_k) \geqslant 0$, for all $x_k \in R^n$, $u_k \in R^m$. This assumption is actually necessary and not at all restrictive because F_k must be a covariance matrix for all x_k, u_k.

We seek a closed-loop optimal feedback controller of the form

$$u_k = C_k(X_k), \quad X_k \triangleq \{x_0,x_1,\dots,x_k\} \qquad (2.3.4)$$

where C_k is a Borel-measurable function of X_k, $k=0,\dots,N-1$.

2.3.2 Solution

The main result is stated as

THEOREM 2.3.1 Define the following quantities:

$$\bar{W}_k(S_{k+1}) \triangleq \tfrac{1}{2} \sum_{i=1}^{n'} tr(S_{k+1}P_k^i)W_k^i \qquad (2.3.5)$$

$$\bar{N}_k(S_{k+1}) \triangleq \tfrac{1}{2} \sum_{i=1}^{n'} tr(S_{k+1}P_k^i)N_k^i \qquad (2.3.6)$$

$$\bar{M}_k(S_{k+1}) \triangleq \tfrac{1}{2} \sum_{i=1}^{n'} tr(S_{k+1}P_k^i)M_k^i \qquad (2.3.7)$$

$$\bar{g}_k(S_{k+1}) \triangleq \tfrac{1}{2} \sum_{i=1}^{n'} tr(S_{k+1}P_k^i)g_k^i \qquad (2.3.8)$$

$$\bar{h}_k(S_{k+1}) \triangleq \tfrac{1}{2} \sum_{i=1}^{n'} tr(S_{k+1}P_k^i)h_k^i \qquad (2.3.9)$$

$$\tilde{R}_k \triangleq R_k + B_k^T S_{k+1} B_k + \bar{M}_k(S_{k+1}) \qquad (2.3.10)$$

$$\tilde{A}_k \triangleq B_k^T S_{k+1} A_k + \bar{N}_k(S_{k+1}) \qquad (2.3.11)$$

$$\tilde{m}_k \triangleq B_k^T S_{k+1} m_k + B_k^T d_{k+1} + \bar{h}_k(S_{k+1}) \qquad (2.3.12)$$

where

$$S_k = Q_k + A_k^T S_{k+1} A_k + \bar{W}_k(S_{k+1}) - \tilde{A}_k \tilde{R}_k^{-1} \tilde{A}_k, \quad S_N = Q_N \qquad (2.3.13)$$

$$d_k = A_k^T S_{k+1} m_k + A_k^T d_{k+1} + \bar{g}_k(S_{k+1}) - \tilde{A}_k \tilde{R}_k^{-1} \tilde{m}_k, \quad d_N = 0 \qquad (2.3.14)$$

$$e_k = \tfrac{1}{2} tr(S_{k+1}P_k^0) + \tfrac{1}{2} m_k^T S_{k+1} m_k + d_{k+1}^T m_k + e_{k+1} - \tfrac{1}{2} \tilde{m}_k^T \tilde{R}_k^{-1} \tilde{m}_k, \quad e_N = 0. \qquad (2.3.15)$$

Then, if

$$\tilde{R}_k > 0, \quad k=0,\ldots,N-1 \tag{2.3.16}$$

the optimal closed-loop controller is linear and is given by

$$u_k = -\tilde{R}_k^{-1}(\tilde{A}_k x_k + \tilde{m}_k), \quad k=0,\ldots,N-1 \tag{2.3.17}$$

and the minimum value of (2.3.1) is

$$\tfrac{1}{2}x_0^T S_o x_0 + d_o^T x_0 + e_o. \tag{2.3.18}$$

PROOF The proof follows by induction and proceeds as follows. First define

$$J_k(X_k) = \min_{u_k,\ldots,u_{N-1}} E_{|X_k} \left[\sum_{i=k}^{N-1} \tfrac{1}{2}(x_i^T Q_i x_i + u_i^T R_i u_i) + \tfrac{1}{2}x_N^T Q_N x_N \right] \tag{2.3.19}$$

so that

$$J_k(X_k) = \min_{u_k} [\tfrac{1}{2}(x_k^T Q_k x_k + u_k^T R_k u_k) + E_{|X_k} J_{k+1}(X_{k+1})] . \tag{2.3.20}$$

It is then easy to show by induction that

$$J_k(X_k) = \tfrac{1}{2}x_k^T S_k x_k + d_k^T x_k + e_k, \quad k=0,\ldots,N. \tag{2.3.21}$$

The next result demonstrates that inequality (2.3.16) holds under certain reasonable conditions on Q_k, R_k and Q_N. In fact these conditions are usually assumed to hold in the standard linear-quadratic control problem formulation.

THEOREM 2.3.2 Suppose that

$$\left.\begin{aligned} Q_N &\geqslant 0 \\ Q_k &\geqslant 0 \\ R_k &> 0 \end{aligned}\right\} \quad k=0,\ldots,N-1 . \tag{2.3.22}$$

Then

$$\tilde{R}_k > 0 \text{ and } S_{k+1} \geqslant 0, \quad k=0,\ldots,N-1. \tag{2.3.23}$$

PROOF The proof is again by induction. Assumption 2.3.3 and equation (2.3.3) imply that if $S_{k+1} \geqslant 0$, then $\bar{M}_{k+1}(S_{k+1}) \geqslant 0$, and inequalities (2.3.22) imply that $\tilde{R}_{N-1} > 0$ and $S_N \geqslant 0$. Now assuming that $S_i \geqslant 0$, $i=k+1,\ldots,N$ we have that $\tilde{R}_{i-1} > 0$, $i=k+1,\ldots,N$. This permits the computation of S_k, d_k and e_k via (2.3.13)-(2.3.15) and allows us to write the equation

$$\tfrac{1}{2}x_k^T S_k x_k + d_k^T x_k + e_k =$$

$$\min_{u_k,\ldots,u_{N-1}} E_{|X_k}\left[\sum_{i=k}^{N-1} \tfrac{1}{2}(x_i^T Q_i x_i + u_i^T R_i u_i) + \tfrac{1}{2}x_N^T Q_N x_N\right]. \tag{2.3.24}$$

Now because of the inequalities (2.3.22), the right-hand side of (2.3.24) is non-negative for all X_k which implies that $S_k \geqslant 0$. This in turn implies that $\tilde{R}_{k-1} > 0$ and so the proof by induction is complete.

2.3.3 Known Special Cases

The standard linear-quadratic-Gaussian problem emerges as a special case if we set

$$f_k(x_k,u_k,\omega_k) = \Gamma_k\omega_k , \tag{2.3.25}$$

with

$$E[\omega_k] = 0, \quad E[\omega_k\omega_k^T] = \Lambda_k . \tag{2.3.26}$$

Here

$$F_k(x_k,u_k) = \Gamma_k\Lambda_k\Gamma_k^T \tag{2.3.27}$$

and it turns out that the optimal feedback controller is independent of $\Gamma_k\Lambda_k\Gamma_k^T$.

The next known case is that of multiplicative noise - see [8], [9] for continuous-time treatment. Here

$$f_k(x_k,u_k,\omega_k) = \sum_{j=1}^{n} (x_k)_j H_k^j\omega_k^1 + \sum_{j=1}^{m} (u_k)_j G_k^j\omega_k^2 \tag{2.3.28}$$

where

$$H_k^j \in R^{n \times q_1}, \quad G_k^j \in R^{n \times q_2} \text{ and } [\omega_k^{1^T} \; \omega_k^{2^T}] = \omega_k^T \tag{2.3.29}$$

with $\omega_k^1 \in R^{q_1}$, $\omega_k^2 \in R^{q_2}$. The j-th component of x_k is denoted by $(x_k)_j$, and

$$E[\omega_k] = 0, \quad E[\omega_k\omega_k^T] = \Lambda_k. \tag{2.3.30}$$

Note that it is not necessary to assume that ω_k^1 and ω_k^2 are uncorrelated although this is often assumed in the literature.

It is not hard to calculate $\bar{W}_k$, $\bar{N}_k$ and $\bar{M}_k$ using (2.3.29) and (2.3.30) and it is clear that the optimal controller depends upon the noise covariance Λ_k.

2.3.4 Novel Special Cases

In this section we display certain novel special cases that have not appeared in the literature hitherto. The first case is that of norm-dependent noise; that is, the dynamic system is described by

$$x_{k+1} = A_k x_k + B_k u_k + m_k + (\tfrac{1}{2}x_k^T D_1 x_k + u_k^T D_2 x_k + \tfrac{1}{2}u_k^T D_3 u_k)^{\frac{1}{2}} . \Gamma_k \omega_k \tag{2.3.31}$$

where $D_1 \geqslant 0$, $D_3 > 0$ and $D_1 - D_2^T D_3^{-1} D_2 \geqslant 0$, and where $E[\omega_k] = 0$, $E[\omega_k \omega_k^T] = \Lambda_k$. Note that if $D_1 = 2I$, $D_2 = D_3 = 0$ we have the noise $\Gamma_k \omega_k$ multiplied by $\| x_k \|$.

For the above case we have

$$\bar{f}_k(x_k, u_k) = 0 \tag{2.3.32}$$

and

$$F_k(x_k, u_k) = (\tfrac{1}{2}x_k^T D_1 x_k + u_k^T D_2 x_k + \tfrac{1}{2}u_k^T D_3 u_k)\Gamma_k \Lambda_k \Gamma_k^T \tag{2.3.33}$$

so that this non-linear system satisfies our assumptions.

Other non-linear examples that satisfy our assumptions are those in which

$$f(x_k, u_k, \omega_k) = \text{sgn}[\phi(x_k, u_k)]\Gamma_k(x_k, u_k)\omega_k \tag{2.3.34}$$

where $\phi : R^{n \times m} \to R^1$ and $\Gamma_k(x_k, u_k)$ is a linear function of x_k, u_k,

$$f(x_k, u_k, \omega_k) = |\beta_k^T x_k + \gamma_k^T u_k| \Gamma_k \omega_k \tag{2.3.35}$$

and

$$f(x_k,u_k,\omega_k) = \|x_k\|\Gamma_k^0\omega_k^0 + \sum_{i=1}^{n} |(x_k)_i|\Gamma_k^i\omega_k^i \qquad (2.3.36)$$

where $\omega_k^0,\ldots,\omega_k^n$ are uncorrelated.

Note that the interesting result that a certain class of non-linear stochastic systems subject to quadratic performance criteria exhibits linear optimal controllers is in contrast to the deterministic non-linear cases (obtained by replacing $\{\omega_k\}$ by a known sequence) for which there do not appear to be closed-form solutions.

2.4 Infinite-time Optimal Control

Solutions of infinite-time optimal control problems can be obtained at least conceptually with the aid of Bellman's partial differential equation. In this section we first prove a well-known but not widely documented theorem on Bellman's equation. We then assume a form for the solution and show that this enables certain non-linear infinite-time problems to be solved. We mention here that Pearson [10] has previously suggested the use of the form of solution of Bellman's equation. However, he failed to state the need for checking a crucial symmetry condition, so that his results, except in the case of his scalar problem, are invalid.

2.4.1 Continuous-time Formulation

We consider the problem of minimizing with respect to $u(\cdot)$ the performance criterion

$$V = \int_0^\infty L(x,u)dt \qquad (2.4.1)$$

where

$$\dot{x} = f(x,u), \quad x(0) = x_0 \tag{2.4.2}$$

where $L:R^{n \times m} \to R^1$, $f:R^{n \times m} \to R^n$, $x \in R^n$, $u \in R^m$.

We make the following assumptions:

ASSUMPTION 2.4.1 The control function $u(\cdot) \in U$, where

$$U \triangleq \{u(\cdot) | u(\cdot) \text{ is piecewise continuous in } t \text{ and } u(t) \in \Omega, \quad t \in [0,\infty)\} \tag{2.4.3}$$

where $\Omega \subseteq R^m$.

ASSUMPTION 2.4.2 The differential equation (2.4.2) has a unique solution defined on the interval $[0,\infty)$ for each $u(\cdot) \in U$.

ASSUMPTION 2.4.3 Along all solutions of (2.4.2), $L(x,u)$ and $f(x,u)$ are piecewise continuous in t, $t \in [0,\infty)$.

ASSUMPTION 2.4.4 We wish to determine a control function $u(\cdot) \in U$ which minimizes (2.4.1) and causes $x(t) \to 0$ as $t \to \infty$.

2.4.2 Bellman-type Theorems

THEOREM 2.4.1 Suppose that there exists a continuously differentiable function $J:R^n \to R^1$ which is positive definite and which satisfies for all $x \in R^n$ the steady-state Bellman equation

$$\min_{u \in \Omega} [L(x,u)+J_x(x)f(x,u)] = 0. \tag{2.4.4}$$

Furthermore, suppose that the solution of $\dot{x} = f(x,u^o(x))$ goes to zero as $t \to \infty$ and that along this solution $u^o(x(\cdot)) \varepsilon U$ where

$$u^o(x) \triangleq \arg \min_{u \varepsilon \Omega} [L(x,u)+J_x(x)f(x,u)]. \tag{2.4.5}$$

Then, (2.4.1) has a minimum in the class of control functions for which $u(\cdot) \varepsilon U$ and which causes $x(t) \to 0$ as $t \to \infty$. The minimum value of V is $J(x_0)$ and the minimizing controller is $u^o(x)$.

PROOF Along trajectories of (2.4.2) we have

$$J(x(t)) - J(x(0)) - \int_0^t \frac{dJ}{d\tau}(x,u)d\tau = 0 \tag{2.4.6}$$

which may be written as

$$J(x(t)) - J(x(0)) - \int_0^t J_x(x)f(x,u)d\tau = 0. \tag{2.4.7}$$

Subtracting this and the other identically zero quantity

$$\int_0^t \min_{u \varepsilon \Omega} [L(x,u)+J_x(x)f(x,u)]\, d\tau \tag{2.4.8}$$

from V yields

$$V = J(x_0) + \int_0^\infty L(x,u)d\tau$$

$$+ \int_0^t \{J_x(x)f(x,u) - \min_{u \varepsilon \Omega} [L(x,u)+J_x(x)f(x,u)]\}d\tau - J(x(t)). \tag{2.4.9}$$

Now, by Assumption 2.4.4 we must consider only those controls

$u(\cdot) \in U$ which cause $x(t) \to 0$ as $t \to \infty$. Consequently restricting attention to this subset of U and noting that as $t \to \infty$, $J(x(t)) \to 0$ (by assumption, J is positive definite) we obtain

$$V = J(x_0) + \int_0^\infty \{ [L(x,u)+J_x(x)f(x,u)] - \min_{u \in \Omega} [L(x,u)+J_x(x)f(x,u)] \} dt. \tag{2.4.10}$$

Clearly the integrand is non-negative and takes on its minimum value when the feasible control $u^o(x)$ is used.

THEOREM 2.4.2 Suppose we assume that $J_x(x) = x^T S(x)$, $S:R^n \to R^{n \times n}$. Then this form is permissible if the matrix $\frac{\partial}{\partial x}[S(x)x]$ is symmetric.

PROOF It is well known that $x^T S(x)$ is the derivative of a scalar function $J(x)$, $J:R^n \to R^1$ if and only if $\frac{\partial}{\partial x}[S(x)x]$ is symmetric.

Using the form $J_x(x) = x^T S(x)$ it is possible to obtain positive definite solutions to certain non-quadratic Bellman equations which are of interest in optimal control. It is worth noting that Pearson [10] has previously used this form for $J_x(x)$, though he suppresses, notationally, the dependence of S on x. However, Pearson fails to check the important condition of symmetry of $\frac{\partial}{\partial x}[S(x)x]$ and this invalidates the results obtained for his examples of dimension two (and higher).

The following theorem combines into one those given above.

THEOREM 2.4.3 Suppose there exists a function $S:R^n \to R^{n \times n}$ such that $\frac{\partial}{\partial x}[S(x)x]$ is symmetric and $\int_0^x y^T S(y)dy$ is positive definite, that satisfies the Bellman equation

$$\min_{u \in \Omega} [L(x,u) + x^T S(x) f(x,u)] = 0. \tag{2.4.11}$$

Furthermore, suppose that the solution of (2.4.2) goes to zero as $t \to \infty$ when using the controller $u^o(x)$ obtained from

$$u^o(x) = \arg \min_{u \in \Omega} [L(x,u) + x^T S(x) f(x,u)]. \tag{2.4.12}$$

Then, $u^o(x(t))$, $t \in [0,\infty)$ minimizes (2.4.1) in the subset of controls $u(\cdot) \in U$ which causes $x(t) \to 0$ as $t \to \infty$, and the minimum value of (2.4.1) is $J(x_o)$.

2.4.3 Certain Non-linear Problems

Consider the bilinear dynamic system

$$\dot{x} = Axu \tag{2.4.13}$$

where $x \in R^n$, $u \in R^1$ and A is a constant $n \times n$ matrix.

The following performance criteria yield Bellman equations that can be solved explicitly:

$$V = \int_0^\infty [\tfrac{1}{8}(x^T Q x)^2 + \tfrac{1}{2}u^2]\, dt, \quad Q > 0, \tag{2.4.14}$$

$$\left.\begin{aligned} V &= \int_0^\infty \tfrac{1}{2} x^T Q x\, dt, \quad Q > 0, \\ \text{where} \quad \Omega &\triangleq \{u \mid |u| \leq 1\}. \end{aligned}\right\} \tag{2.4.15}$$

Taking (2.4.14) the Bellman equation is

$$\min_u [\tfrac{1}{8}(x^T Q x)^2 + \tfrac{1}{2}u^2 + x^T S(x) A x u] = 0. \tag{2.4.16}$$

Assuming, for convenience, that S(x) is symmetric yields

$$u^o(x) = -\tfrac{1}{2}x^T[S(x)A+A^TS(x)]x \qquad (2.4.17)$$

and

$$\frac{1}{8}(x^TQx)^2 - \frac{1}{8}\{x^T[S(x)A+A^TS(x)]x\}^2 = 0. \qquad (2.4.18)$$

This equation is satisfied for all x if S (a constant matrix) is chosen to satisfy

$$Q \pm [SA+A^TS] = 0. \qquad (2.4.19)$$

If the real parts of the eigenvalues of A are all negative (positive) then $Q + (-)[SA+A^TS] = 0$ yields a unique positive definite solution. Hence $J(x) = \tfrac{1}{2}x^TSx$ and $u^o = +x^TQx$ ($u^o = -x^TQx$).

If (2.4.15) is minimized the appropriate Bellman equation is

$$\min_{u \in \Omega} [\tfrac{1}{2}x^TQx+x^TS(x)Axu] = 0 \qquad (2.4.20)$$

which yields

$$u^o(x) = -\operatorname{sign} x^T[S(x)A+A^TS(x)]x \qquad (2.4.21)$$

and

$$\tfrac{1}{2}x^TQx - \tfrac{1}{2}|x^T[S(x)A+A^TS(x)]x| = 0. \qquad (2.4.22)$$

Here

$$u^o(x) = \operatorname{sign} x^TQx = +1 \qquad (2.4.23)$$

if the real parts of the eigenvalues of A are all negative, and

$$u^o(x) = -1 \qquad (2.4.24)$$

if they are all positive.

Note that in both cases a quadratic $J(x)$ solves Bellman's equation.

2.4.4 Non-quadratic Performance Criterion

Suppose that the system to be controlled is

$$\dot{x} = Ax + Bu \tag{2.4.25}$$

where the constant pair (A,B) is completely controllable. The performance criterion is

$$V = \int_0^\infty \tfrac{1}{2}\{x^TQx(1+x^TPx)+(x^TPx)x^TPBR^{-1}B^TPx(1+x^TPx) + u^TRu\}dt \tag{2.4.26}$$

where $Q > 0$, $R > 0$ and P is the unique positive definite symmetric solution of

$$Q + PA + A^TP - PBR^{-1}B^TP = 0. \tag{2.4.27}$$

It is easy to check that the choice

$$S(x) = (1+x^TPx)P \tag{2.4.28}$$

results in

$$J(x) = \tfrac{1}{4}[(1+x^TPx)^2-1] \tag{2.4.29}$$

and

$$u^o(x) = -(1+x^TPx)R^{-1}B^TPx \tag{2.4.30}$$

and that all solutions of $\dot{x} = Ax + Bu^o(x)$ go to zero as $t \to \infty$. Note lastly that

$$\frac{\partial}{\partial x}[S(x)x] = (1+x^TPx)P + 2Pxx^TP \qquad (2.4.31)$$

which is symmetric.

For small x, $u^o(x) \approx -R^{-1}B^TPx$ which is the standard controller of linear-quadratic theory, while for large x, $u^o(x) \approx -x^TPxR^{-1}B^TPx$ which is cubic in x.

We refer the reader to Speyer [11] for an extension of this example to a stochastic setting.

2.5 Systems Homogeneous-in-the-input

This is a class of systems which permits a complete analysis of its stabilizability properties and also yields explicit closed-loop controllers which minimize a wide variety of performance criteria [12].

2.5.1 Formulation

Specifically, a system is said to be homogeneous-in-the-input if it is of the form

$$\dot{x} = \sum_{i=1}^{m} B_i(x)u_i, \quad x(0) = x_0 \qquad (2.5.1)$$

where $x \in R^n$, $u_i \in R^1$, $i=1,\ldots,m$, and where the $B_i(x)$, $B_i: R^n \to R^n$ are assumed to be continuous functions of x. We shall further assume that the controls are chosen in feedback form as $u_i(x) = g_i(x)$, $i=1,\ldots,m$ where, unless otherwise stated, $g_i: R^n \to R^1$ is a continuous function of x.

2.5.2 Stabilizability

In this section we provide necessary conditions and sufficient conditions for stabilizability of (2.5.1). The gap between these conditions is small.

THEOREM 2.5.1 Suppose that $u_i(x) = g_i(x)$, $i=1,\ldots,m$ and that g_i and B_i are once continuously differentiable with respect to x. A necessary condition for x = 0 to be an asymptotically stable equilibrium point of (2.5.1) is that there exists a positive definite function V(x), $V:R^n \to R^1$, which is continuously differentiable, such that there is no non-zero $x \in R^n$ for which

$$V_x(x)B_i(x), \quad i=1,\ldots,m \tag{2.5.2}$$

are all zero.

PROOF As the right-hand side of (2.5.1) with the above choice of controls is independent of t and is continuously differentiable with respect to x, an inverse theorem of Liapunov [13] applies direct. This theorem guarantees the existence of a positive definite V(x) with $\dot{V}(x) = V_x(x)\dot{x}$ negative definite. Now, along trajectories of (2.5.1)

$$\dot{V}(x) = \sum_{i=1}^{m} V_x(x)B_i(x)g_i(x) \tag{2.5.3}$$

and it follows immediately that this expression can be negative definite only if the theorem is true.

The following theorem provides sufficient conditions for stabilizability.

THEOREM 2.5.2 A sufficient condition for x = 0 to be made

into an asymptotically stable equilibrium point of (2.5.1) is that there exists a positive definite function $V(x)$, $V:R^n \to R^1$, which is once continuously differentiable, such that there is no non-zero $x \in R^n$ for which

$$V_x(x)B_i(x), \quad i=1,\dots,m \tag{2.5.4}$$

are all zero. If in addition $V(x)$ is radially unbounded, then $u_i(x) = -V_x(x)B_i(x)$ causes $x = 0$ to be a globally asymptotically stable equilibrium point.

PROOF Let us set $u_i(x) = -V_x(x)B_i(x)$, $i=1,\dots,m$. Then, the $u_i(x)$ are continuous and

$$\dot{V}(x) = - \sum_{i=1}^{m} [V_x(x)B_i(x)]^2 < 0, \quad x \neq 0. \tag{2.5.5}$$

The proof now follows from a standard Liapunov theorem [13].

2.5.3 A Bilinear Example

Theorem 2.5.2 is especially useful in those cases where (2.5.1) cannot be stabilized by means of constant controls $u_i = k_i$, $i=1,\dots,m$. For example, let

$$B_1 = \begin{pmatrix} 0 & 1 \\ 1 & 0 \end{pmatrix}, \quad B_2 = \begin{pmatrix} -1 & 0 \\ 0 & 1 \end{pmatrix} \tag{2.5.6}$$

and

$$\dot{x} = B_1xu_1 + B_2xu_2. \tag{2.5.7}$$

In this case it is easy to see that there are no real numbers

k_1 and k_2 such that $\sum_{i=1}^{m} B_i k_i$ is a stability matrix. In other words, constant controls cannot stabilize (2.5.7). However, if we let $V(x) = \frac{1}{2}x^T x$ we see that

$$V_x(x)B_1 x = 2x_1 x_2 \tag{2.5.8}$$

and

$$V_x(x)B_2 x = - x_1^2 + x_2^2 \tag{2.5.9}$$

which do not vanish simultaneously for any non-zero $x \in R^2$. As $V(x)$ is radially unbounded we have the result that

$$u_1(x) = - 2x_1 x_2 \tag{2.5.10}$$

$$u_2(x) = x_1^2 - x_2^2 \tag{2.5.11}$$

globally asymptotically stabilize (2.5.7).

Note that it is a trivial matter to prove that if there exist $k_i \in R^1$, $i=1,\ldots,m$ such that $\sum_{i=1}^{m} B_i k_i$ is a stability matrix, then there exists $V(x) = \frac{1}{2}x^T Sx$, $S > 0$, such that the scalars $V_x(x)B_i x = \frac{1}{2}x^T(SB_i + B_i^T S)x$, $i=1,\ldots,m$ are not zero simultaneously for a non-zero $x \in R^n$.

2.5.4 Optimal Control

In Section 2.5.2 we showed that subject to very reasonable conditions, $u_i(x) = - V_x(x)B_i(x)$, $i=1,\ldots,m$ asymptotically stabilize (2.5.1). It follows easily that

$$u_i(x) = - [V_x(x)B_i(x)]^{\frac{1}{2p+1}}, \quad i=1,\ldots,m \tag{2.5.12}$$

where p is a non-negative integer, also accomplishes this. This raises the question of whether or not an 'optimal' stabilizing controller can be designed. It turns out that the answer to this question is in the affirmative and that stabilizing controllers which minimize a wide variety of performance criteria can be constructed for (2.5.1).

We define the performance criterion

$$\bar{V} = \int_{0}^{\infty}\{q(x) + \frac{1}{2(p+1)} \sum_{i=1}^{m} u_i^{2(p+1)}\}dt \qquad (2.5.13)$$

where $q(x)$ is a positive definite function of $x \in R^n$ and p is a non-negative integer. We then have the following theorem.

THEOREM 2.5.1 Suppose that there exists a radially unbounded, positive definite function $V(x)$ which is once continuously differentiable, which satisfies the Bellman equation

$$\min_{u_1,\ldots,u_m} [q(x)+ \frac{1}{2(p+1)} \sum_{i=1}^{m} u_i^{2(p+1)} + \sum_{i=1}^{m} V_x(x)B_i(x)u_i] = 0. \qquad (2.5.14)$$

Then, the controls (2.5.12) globally asymptotically stabilize (2.5.1) and minimize (2.5.13) in the class of control functions which causes $x(t) \to 0$ as $t \to \infty$.

PROOF Carrying out the minimization in (2.5.14) yields (2.5.12) which, when substituted back into (2.5.14) yields

$$q(x) - \frac{2p+1}{2(p+1)} \sum_{i=1}^{m} [V_x(x)B_i(x)]^{\frac{2(p+1)}{2p+1}} = 0. \qquad (2.5.15)$$

Note that in order for (2.5.15) to hold, $V_x(x)B_i(x)$, $i=1,\ldots,m$ cannot all be zero for a non-zero $x \in R^n$ - this is precisely a sufficient condition for the controller (2.5.12) to be globally asymptotically stabilizing. An application of Theorem 2.4.1 then completes the proof.

Because of the special form of the Bellman equation we can, without loss of generality, replace q(x) in (2.5.13) by

$$\frac{2p+1}{2(p+1)} \sum_{i=1}^{m} [V_x(x)B_i(x)]^{\frac{2(p+1)}{2p+1}} . \tag{2.5.16}$$

In other words, instead of specifying q(x) and solving the Bellman equation for V(x) we can choose a suitable V(x) and hence specify q(x). The next theorem is a statement of this approach.

THEOREM 2.5.2 Suppose there exists a radially unbounded, positive definite function V(x), $V: R^n \to R^1$, which is once continuously differentiable, such that there is no non-zero $x \in R^n$ for which $V_x(x)B_i(x)$, $i=1,\ldots,m$ are all zero. Then, the controls (2.5.12) globally asymptotically stabilize (2.5.1) and minimize (2.5.13) in the class of control functions which causes $x(t) \to 0$ as $t \to \infty$ where q(x) is given by (2.5.15).

An appropriate choice of V(x) will yield a desired q(x). In many cases the class of q(x) defined by (2.5.15) with V(x) a quadratic function, is adequate. Indeed, returning to the example of Section 2.5.3 we have, for p = 0,

$$\begin{aligned} q(x) &= \sum_{i=1}^{2} \tfrac{1}{2}[x^T B_i x]^2 = \tfrac{1}{2}[(2x_1x_2)^2+(x_1^2-x_2^2)^2] \\ &= \tfrac{1}{2}(x_1^2+x_2^2)^2. \end{aligned} \tag{2.5.17}$$

The control strategies (2.5.10), (2.5.11) therefore minimize

$$\bar{V} = \int_0^\infty \{\tfrac{1}{2}(x_1^2+x_2^2)^2+\tfrac{1}{2}(u_1^2+u_2^2)\}dt. \qquad (2.5.18)$$

Note that for large p, (2.5.13) is approximately

$$\bar{V} \approx \int_0^\infty \{ \sum_{i=1}^m |V_x(x)B_i(x)| + \frac{1}{2(p+1)} \sum_{i=1}^m u_i^{2(p+1)}\}dt \qquad (2.5.19)$$

which leads us to the following result.

THEOREM 2.5.3 Suppose that there exists a radially unbounded, positive definite function $V(x)$, $V:R^n \to R^1$, which is once continuously differentiable such that there is no non-zero $x \in R^n$ for which $V_x(x)B_i(x)$, $i=1,\ldots,m$ are all zero. Then, the controls

$$u_i(x) = - \text{sign}[V_x(x)B_i(x)], \quad i=1,\ldots,m \qquad (2.5.20)$$

globally asymptotically stabilize (2.5.1) and minimize

$$J = \int_0^\infty q(x)dt \qquad (2.5.21)$$

in the class of control functions which causes $x(t) \to 0$ as $t \to \infty$ and which satisfies the control constraints

$$-1 \leqslant u_i \leqslant 1, \quad i=1,\ldots,m \qquad (2.5.22)$$

where, with no loss of generality

$$q(x) = \sum_{i=1}^m |V_x(x)B_i(x)|. \qquad (2.5.23)$$

Finally, we mention that it is not hard to show that $u_i(x) = -[V_x(x)B_i(x)]^{2p+1}$, $i=1,\ldots,m$ globally asymptotically stabilizes (2.5.1) and minimizes

$$\bar{V} = \int_0^\infty \{q(x) + \frac{2p+1}{2(p+1)} \sum_{i=1}^m |u_i|^{\frac{2(p+1)}{2p+1}}\}dt \qquad (2.5.24)$$

where, with no loss of generality

$$q(x) = \frac{1}{2(p+1)} \sum_{i=1}^m [V_x(x)B_i(x)]^{2(p+1)}. \qquad (2.5.25)$$

2.5.5 Non-homogeneous Extension

We extend the results of Section 2.5.4 to systems of the type

$$\dot{x} = f(x) + \sum_{i=1}^m B_i(x)u_i, \quad x(0) = x_0. \qquad (2.5.26)$$

THEOREM 2.5.4 Suppose that there exists a radially unbounded, positive definite function $V(x)$, $V:R^n \to R^1$, which is once continuously differentiable, such that $V_x(x)f(x)$ is negative semi-definite. Suppose further that there is no non-zero $x \in R^n$ for which $V_x(x)f(x)$ and $V_x(x)B_i(x)$, $i=1,\ldots,m$ are all zero.

Then,

$$u_i(x) = -V_x(x)B_i(x), \quad i=1,\ldots,m \qquad (2.5.27)$$

globally asymptotically stabilizes (2.5.26) and minimizes

$$\bar{V} = \int_0^\infty \{q(x)+\tfrac{1}{2}\sum_{i=1}^m u_i^2\}dt \qquad (2.5.28)$$

in the class of control functions which causes $x(t) \to 0$ as $t \to \infty$. Here $q(x)$ is, without loss of generality, given by the positive definite expression

$$- V_x(x)f(x) + \tfrac{1}{2}\sum_{i=1}^{m} [V_x(x)B_i(x)]^2. \qquad (2.5.29)$$

Note that there is often considerable flexibility in choosing a V function to obtain a suitable q(x). Indeed in many cases a quadratic function is adequate. For example, in the case

$$\dot{x} = Ax + \sum_{i=1}^{m} B_i x u_i \qquad (2.5.30)$$

where A has eigenvalues with non-positive real parts, there exists an S such that $V(x) = \tfrac{1}{2}x^T Sx$ satisfies the conditions of the theorem. It then turns out that

$$u_i(x) = -\tfrac{1}{2}x^T(SB_i+B_i^T S)x, \quad i=1,\dots,m \qquad (2.5.31)$$

minimizes

$$J = \int_0^\infty \{\tfrac{1}{2}x^T Qx+\tfrac{1}{2}\sum_{i=1}^{m} [x^T(SB_i+B_i^T S)x]^2+\tfrac{1}{2}\sum_{i=1}^{m} u_i^2\}dt \qquad (2.5.32)$$

where $Q = -(SA+A^T S) \geqslant 0$.

2.6 Optimal Control of Quadratic Systems

2.6.1 Formulation

We consider here the optimal control of systems of the type

$$\frac{d}{dt} f_i(x) = \tfrac{1}{2}x^T A_i x + x^T B_i u, \quad x_i(0) = x_{0_i}, \quad i=1,\dots,n \qquad (2.6.1)$$

where $f_i:R^n \to R^1$ is a continuous function, $A_i \in R^{n\times n}$, $B_i \in R^{n\times m}$, i=1,...,n. We assume that (2.6.1) has a solution for each piecewise continuous control function. The special case where $f_i = x_i$, was first treated in [14], while the above more general formulation was suggested by W.M. Getz.

The control problem is to design a controller for (2.6.1)

which minimizes the performance criterion

$$V = \int_0^\infty \tfrac{1}{2}(x^T Qx+u^T Ru)dt \qquad (2.6.2)$$

where $Q \varepsilon R^{n \times n}$, $R \varepsilon R^{m \times m}$ are symmetric, positive definite matrices. Note that without loss of generality A_i can be assumed to be symmetric.

An alternative, finite-time, problem is to design a controller which minimizes

$$V = \int_0^T \tfrac{1}{2}(x^T Qx+u^T Ru)dt \qquad (2.6.3)$$

subject to the terminal constraint

$$x(T) = b \neq 0 \qquad (2.6.4)$$

where T is given.

2.6.2 Solutions

We first state as theorems the solutions of the above two problems. We then prove one of these theorems.

THEOREM 2.6.1 Suppose that there exists a vector $c \varepsilon R^n$ which satisfies the Riccati-like equation

$$Q + \sum_{i=1}^{n} c_i A_i - (\sum_{i=1}^{n} c_i B_i) R^{-1} (\sum_{i=1}^{n} c_i B_i)^T = 0. \qquad (2.6.5)$$

Suppose, further, that each solution of the differential equations

$$\frac{d}{dt} f_i(x) = \tfrac{1}{2}x^T A_i x - x^T B_i R^{-1}\left(\sum_{i=1}^{n} c_i B_i\right)^T x, \quad x_i(0) = x_{0_i} \tag{2.6.6}$$

$i=1,\ldots,n$, is defined for $t \in [0,\infty)$ and that $x(t) \to 0$ as $t \to \infty$. Then, the minimum value of (2.6.2) in the class of controls which causes $x(t) \to 0$ as $t \to \infty$ is

$$J = c^T[f(x_0)-f(0)] \tag{2.6.7}$$

and the optimal controller which achieves this minimum is

$$u = -R^{-1}\left(\sum_{i=1}^{n} c_i B_i\right)^T x. \tag{2.6.8}$$

THEOREM 2.6.2 Suppose that there exists a vector $c \in R^n$ which satisfies (2.6.5) and which causes each solution of (2.6.6) to have the value b at $t = T$; i.e. $x(T) = b$. Then, the minimum value of (2.6.3) is given by

$$J = c^T[f(x_0)-f(b)] \tag{2.6.9}$$

and the optimal controller which achieves this minimum value and which causes $x(T) = b$ is

$$u = -R^{-1}\left(\sum_{i=1}^{n} c_i B_i\right)^T x. \tag{2.6.10}$$

As the proofs of the two theorems are very similar we prove only Theorem 2.6.1.

PROOF OF THEOREM 2.6.1 Along a solution of (2.6.1) we have that

$$c^T f(x_0) - c^T f(x(t)) + \int_0^t \frac{d}{d\tau}[c^T f(x(\tau))]\,d\tau = 0. \tag{2.6.11}$$

This, suitably rewritten, is

$$c^T f(x_0) - c^T f(x(t)) + \int_0^t (\tfrac{1}{2}x^T \sum_{i=1}^n c_i A_i x + x^T \sum_{i=1}^n c_i B_i u)\,d\tau = 0. \tag{2.6.12}$$

Adding this identically zero quantity to (2.6.2) and letting $t \to \infty$ yields

$$\begin{aligned} V = {} & c^T f(x_0) - \lim_{t \to \infty} c^T f(x(t)) \\ & + \int_0^\infty \tfrac{1}{2}\{x^T [Q + \sum_{i=1}^n c_i A_i - (\sum_{i=1}^n c_i B_i) R^{-1} (\sum_{i=1}^n c_i B_i)^T] x \\ & + [u + R^{-1}(\sum_{i=1}^n c_i B_i)^T x]^T R [u + R^{-1}(\sum_{i=1}^n c_i B_i)^T x]\}\,dt \end{aligned} \tag{2.6.13}$$

which, upon using (2.6.5) becomes

$$\begin{aligned} V = {} & c^T f(x_0) - \lim_{t \to \infty} c^T f(x(t)) \\ & + \int_0^\infty \tfrac{1}{2} [u + R^{-1}(\sum_{i=1}^n c_i B_i)^T x]^T R [u + R^{-1}(\sum_{i=1}^n c_i B_i)^T x]\,dt. \end{aligned} \tag{2.6.14}$$

As we only draw our controls from that class which causes $x(t) \to 0$ as $t \to \infty$ we see that because of the continuity of f, $\lim_{t \to \infty} c^T f(x(t)) = c^T f(0)$. Now, by assumption, (2.6.8) yields a control function which causes $x(t) \to 0$ as $t \to \infty$ so that, by inspection, (2.6.8) minimizes (2.6.14) and the minimum value of (2.6.14) is given by (2.6.7).

Although the restrictions of the theorems, particularly the solvability of (2.6.5), are quite stringent we nevertheless have exhibited here a class of highly non-linear systems whose optimal controllers relative to quadratic performance criteria, are linear. Note that except for the stability condition on (2.6.6), the optimal controller is independent of the function $f(x)$.

2.7 Conclusion

In this chapter we covered a broad class of non-linear-quadratic control problems. Sections 2.1 and 2.2 dealt with the minimization of exponential performance criteria subject to linear dynamic systems, in a stochastic setting. We showed that the optimal controller is linear but that, in the presence of measurement noise, it may be infinite-dimensional. In the case of no measurement noise we demonstrated a close tie between the stochastic control problem and a class of deterministic differential games. One of the games, the non-cooperative one, is often used to design a type of 'worst-case controller' for linear systems. The tie between this game and the stochastic control problem seems to imply that this type of design is not as conservative as its name would imply.

We note here that exponential performance criteria have implications also for fuzzy set theory [15]. In fact the exponential criteria can be easily interpreted as membership functions of fuzzy sets and this determines that the notion of 'confluence of fuzzy goals and constraints', introduced by Bellman and Zadeh, be interpreted in novel ways.

Section 2.3 was concerned with non-linear stochastic systems of special structure. The formulation includes control and state-dependent noise in linear systems, which have received attention elsewhere. Our general formulation yields a number of interesting further special cases. We remark that again here the optimal feedback controller is linear and depends upon the statistics of the system noise.

Section 2.4 treated infinite-time control problems via the Bellman equation. This allowed the solution of some special non-linear control problems and paved the way for Section 2.5 which was devoted to the stabilizability and optimal control of non-linear systems homogeneous-in-the-input. This is one of the few classes of non-linear systems which allow the determination of 'closed form' non-linear feedback controllers.

Finally, Section 2.6 was devoted to the optimal control of a class of highly non-linear systems. The results, though rather restrictive, provide insight into quadratic and more non-linear systems.

In the next Chapter we turn to matrix theory - in particular the properties of copositive and related matrices which play a role in the understanding of non-convex quadratic forms are studied. Certain matrix-theoretic results are then exploited in obtaining conditions for the solutions of quadratic differential equations to have finite escape times.

2.8 References

[1] JACOBSON, D.H. Optimal Stochastic Linear Systems with Exponential Performance Criteria and their Relation to Deterministic Differential Games. IEEE Trans. Automatic Control, AC-18, 1973, pp. 124-131.

[2] JACOBSON, D.H. On a Result in Stochastic Optimal Control. IEEE Trans. Automatic Control, AC-18, 1973, pp. 411-412.

[3] SPEYER, J.L., DEYST, J. & JACOBSON, D.H. Optimization of Stochastic Linear Systems with Additive Measurement and Process Noise Using Exponential Performance Criteria. IEEE Trans. Automatic Control, AC-19, 1974, pp. 358-366.

[4] IEEE Trans. Automatic Control, AC-16, December 1971. Special issue on the Linear-Quadratic-Gaussian Problem.

[5] WITSENHAUSEN, H.S. Separation of Estimation and Control for Discrete-Time Systems. Proc. IEEE, 59, 1971, pp. 1557-1566.

[6] SPEYER, J.L. An Adaptive Terminal Guidance Scheme Based on an Exponential Cost Criterion with Application to Homing Missile Guidance. IEEE Trans. Automatic Control, AC-21, 1976, pp. 371-375.

[7] JACOBSON, D.H. A General Result in Stochastic Optimal Control of Non-linear Discrete-Time Systems with Quadratic Performance Criteria. J. Math. Anal. Appl., 47, 1974, pp. 153-161.

[8] WONHAM, W.M. Optimal Stationary Control of a Linear System with State-dependent Noise. SIAM J. Control, 5, 1967, pp. 486-500.

[9] McLANE, P.J. Optimal Stochastic Control of Linear Systems with State and Control Dependent Disturbances. IEEE Trans. Automatic Control, AC-16, 1971, pp. 793-798.

[10] PEARSON, J.D. Approximation Methods in Optimal Control I. Sub-optimal Control. J. Electronics and Control, 13, 1962, pp. 453-469.

[11] SPEYER, J.L. A Non-linear Control Law for a Stochastic Infinite Time Problem. IEEE Trans. Automatic Control, AC-21, 1976, pp. 560-564.

[12] JACOBSON, D.H. Stabilization and Optimal Control of Non-linear Systems Homogeneous-in-the-input. In Proceedings of a Conference on Directions in Decentralized Control, Many-Person Optimization and Large-Scale Systems, held in Boston, Mass., 1-3 September 1975 and published by Plenum Press, New York, 1976, pp. 389-399.

[13] HAHN, W. Stability of Motion. Springer Verlag, Berlin, Heidelberg, New York, 1967.

[14] JACOBSON, D.H. On the Optimal Control of Systems of Quadratic and Bilinear Differential Equations. Proc. 6th IFAC World Congress, Boston, Mass., August, 1975.

[15] JACOBSON, D.H. On Fuzzy Goals and Maximizing Decisions in Stochastic Optimal Control. J. Math. Anal. Appl., 55, 1976, pp. 434-440.

3. COPOSITIVE MATRICES, NON-CONVEX QUADRATIC FORMS AND QUADRATIC DIFFERENTIAL EQUATIONS

3.1 Introduction to Copositive Matrices

It turns out that much of the theory of control, optimization and stability depends heavily upon matrix theory - in particular upon eigenvalue theory for general square matrices and upon the properties of quadratic forms. Usually one assumes that quadratic forms are positive semi-definite, that is, $x^TQx \geqslant 0$ for all $x \in R^n$. It is trivial to show that positive semi-definite quadratic forms are convex functions (the converse is also true), and this is an added attraction for their use. Furthermore, the Sylvester test for positive definite matrices is widely known and relatively easy to apply, at least in a few dimensions. Last but not least, Liapunov stability theory is based solidly upon the notion of a positive definite (quadratic) form.

Though useful, the notion of a positive semi-definite matrix (or quadratic form) is very restrictive. For example, if Q is positive semi-definite we have a fortiori that $x^TQx \geqslant 0$ for all $x \geqslant 0$. However, the set of matrices for which $x^TQx \geqslant 0$ for all $x \geqslant 0$ is larger than the set of positive semi-definite matrices. This class, which is called the class of <u>copositive</u> matrices, is clearly important in problems of constrained minimization such as that of minimizing

$$x^TQx \tag{3.1.1}$$

subject to

$$Ax \geqslant b \tag{3.1.2}$$

where $x \in R^n$.

The two major early works on copositive matrices are, in our opinion, [1] and [2]. As pointed out above, any $n \times n$ positive semi-definite matrix is also copositive as is any $n \times n$ matrix having non-negative elements. In [1] Diananda therefore conjectured that the class of copositive matrices consists of matrices which are sums of positive semi-definite matrices and matrices having non-negative elements. In fact, Diananda was able to prove that this is true for $n \leqslant 4$ (recall that $Q \in R^{n \times n}$), while A. Horn (see [1], [3]) supplied a counter-example when $n = 5$. Gaddum's approach [2] to the problem was different; he related the copositive quadratic forms to linear inequalities which allowed him to deduce necessary and sufficient conditions for copositivity. These conditions did not yield much insight into the structure of copositive matrices when $n \geqslant 5$, but this aspect has since been studied [3] - [6].

In the next sections we present in a little detail the major results of Diananda, Gaddum and others and then extend and use them in non-convex quadratic programming and quadratic differential equation theory.

3.1.1 Properties of Copositive Matrices

DEFINITION 3.1.1 A symmetric matrix $Q \in R^{n \times n}$ is said to be copositive if $x^T Qx \geqslant 0$ for all $x \in R^n$ such that $x \geqslant 0$.

Clearly all positive semi-definite matrices are copositive as are all matrices with non-negative elements. Though the conjecture that all copositive matrices can be written as sums of these matrices is false, we do have the following theorem.

THEOREM 3.1.1 (Diananda) Every copositive matrix $Q \in R^{n \times n}$ can be written as the sum of a positive semi-definite matrix $S \in R^{n \times n}$ and a matrix $P \in R^{n \times n}$ having non-negative elements, if and only if $n \leqslant 4$.

PROOF The result is obvious when n = 1 and is more cumbersome than difficult to prove when n = 3, 4. We refer the reader to Diananda [1] for details. The case n = 2 is presented here by way of illustration.

First we note that $x^T Qx = q_{11}x_1^2 + 2q_{12}x_1x_2 + q_{22}x_2^2$, which implies that we must have $q_{11} \geqslant 0$, $q_{22} \geqslant 0$. If q_{12} is non-negative we have that Q has all its elements non-negative. On the other hand, if q_{12} is negative, both q_{11} and q_{22} must be positive, so that $x_1 = -\frac{q_{12}}{q_{11}} x_2$, $x_2 > 0$ are both defined and are positive. Substituting these values into $x^T Qx$ yields

$$\frac{q_{12}^2}{q_{11}} x_2^2 - 2\frac{q_{12}^2}{q_{11}} x_2^2 + q_{22}x_2^2 \qquad (3.1.3)$$

which is non-negative for all $x_2 > 0$ if and only if

$$q_{11}q_{22} - q_{12}^2 \geqslant 0. \qquad (3.1.4)$$

Inequality (3.1.4) together with $q_{11} > 0$, $q_{22} > 0$ implies that Q is positive semi-definite, and the theorem is proved.

Diananda's conjecture that every copositive Q could be written as

$$Q = S + P \qquad (3.1.5)$$

for all n was proved to be false by A. Horn's counter-example in five dimensions

$$x^TQx = (x_1+x_2+x_3+x_4+x_5)^2 - 4x_1x_2 - 4x_2x_3 - 4x_3x_4 - 4x_4x_5 - 4x_5x_1 \ . \qquad (3.1.6)$$

This function can be written as

$$(x_1-x_2+x_3+x_4-x_5)^2 + 4x_2x_4 + 4x_3(x_5-x_4) \qquad (3.1.7)$$

which is clearly non-negative when $x_i \geqslant 0$, i=1,...,5 and $x_5 \geqslant x_4$, or

$$(x_1-x_2+x_3-x_4+x_5)^2 + 4x_2x_5 + 4x_1(x_4-x_5) \qquad (3.1.8)$$

which is non-negative when $x_i \geqslant 0$, i=1,...,5 and $x_4 \geqslant x_5$. Furthermore, this copositive quadratic form can be shown to be an extreme form [3]. As Q is not positive semi-definite and as it does not have all its elements non-negative, it then follows that Q cannot be written as S + P.

A result dual to that given in Theorem 3.1.1 is stated in Theorem 3.1.2 which follows the next definition.

DEFINITION 3.1.2 A symmetric matrix $B \in R^{n \times n}$ is said to be completely positive if it can be factored in the form $B = F^TF$ where $F \in R^{m \times n}$, m a positive integer, and where $F_{ij} \geqslant 0$ for all i,j.

THEOREM 3.1.2 Every symmetric positive semi-definite matrix in $R^{n \times n}$ having non-negative elements is completely positive if and only if $n \leqslant 4$.

PROOF The set of copositive matrices Q contains the set of matrices which are sums of positive semi-definite matrices S and matrices with non-negative elements P, i.e.

$$Q \supseteq P + S. \tag{3.1.9}$$

Furthermore, the spaces of copositive quadratic forms and completely positive forms are both closed convex cones and each is the dual cone of the other [3]. The properties of duals then imply that

$$B = Q^+ \subseteq (P+S)^+ = P^+ \cap S^+ = P \cap S \tag{3.1.10}$$

where '+' denotes 'dual' and where the last equality holds because both P and S are self-dual. The theorem now follows from Theorem 3.1.1 and the fact that equality in (3.1.9) implies equality in (3.1.10), and vice versa.

An example which illustrates that Theorem 3.1.2 cannot be extended to $n \geqslant 5$ is

$$\begin{aligned} x^T Bx &= x_1^2 + x_2^2 + x_3^2 + x_4^2 + x_5^2 \\ &\quad + x_1x_2 + x_1x_5 + x_2x_3 + \tfrac{3}{2}\, x_3x_4 + x_4x_5 \\ &= (x_2+\tfrac{1}{2}x_1+\tfrac{1}{2}x_3)^2 + (x_5+\tfrac{1}{2}x_1+\tfrac{1}{2}x_4)^2 \\ &\quad + \tfrac{1}{2}(x_1-\tfrac{1}{2}x_3-\tfrac{1}{2}x_4)^2 + \tfrac{5}{8}(x_3+x_4)^2 \end{aligned} \tag{3.1.11}$$

which is fully discussed in [3].

We turn now to Gaddum's work [2] where we require the following definition.

DEFINITION 3.1.3 Suppose the variables $x_{k_1}, x_{k_2}, \ldots, x_{k_s}$ are set to zero. Then the quadratic form $x^T Q x$ becomes

$$\sum_{i,j \neq k_1,\ldots,k_s} q_{ij} x_i x_j \qquad (3.1.12)$$

which may be written in terms of the not necessarily zero components of x, referred to as, say $\bar{x}$, as

$$\bar{x}^T Q^{k_1 \ldots k_s} \bar{x} \ . \qquad (3.1.13)$$

We define $Q^{k_1 \ldots k_s}$ to be a principal minor of Q.
When we write $Q^{k_1 \ldots k_s} x$ in the sequel we shall assume that x has the appropriate number of components.

For our purposes, the main theorem of Gaddum's work is

THEOREM 3.1.3 A necessary and sufficient condition that $x^T Q x$ be copositive is that for each principal minor $Q^{k_1 \ldots k_s}$ of Q, the system $Q^{k_1 \ldots k_s} x \geqslant 0$, $x \geqslant 0$ has a non-trivial solution.

Gaddum's theorem is useful in that it relates the properties of a non-convex quadratic form to those of systems of linear inequalities. Such systems of equations may be tested for solutions via a linear program as the following theorem indicates.

THEOREM 3.1.4 The system of linear inequalities $Q^{k_1 \ldots k_s} x \geq 0$, $x \geq 0$ has a non-trivial solution if and only if the following linear program has a non-trivial solution $(\hat{x}, \hat{\varepsilon})$ with $\hat{x} \neq 0$ and $\hat{\varepsilon} \geq 0$:

$$\max_{x,\varepsilon} \varepsilon \tag{3.1.14}$$

subject to the constraints

$$Q^{k_1 \ldots k_s} x \geq \varepsilon e \tag{3.1.15}$$

$$e \geq x \geq 0 \tag{3.1.16}$$

where e is a vector of ones of the same dimension as x.

PROOF If the linear inequalities have a non-trivial solution $\bar{x}$ this qualifies as a starting solution for the linear program, with $\bar{\varepsilon} \geq 0$. Consequently, the linear program has a non-trivial solution. On the other hand, if the linear program has a non-trivial solution, so do also the linear inequalities.

3.2 Extensions

In this section we extend certain of the above results. We begin by generalizing the constraint set from $x \geq 0$ to $Ax \geq 0$.

3.2.1 General Linear Constraints

We here characterize matrices Q for which $x^T Q x \geq 0$ for all x such that $Ax \geq 0$, $A \in R^{m \times n}$.

THEOREM 3.2.1 Suppose that $x^T Q x \geq 0$ for all x such that $Ax \geq 0$ where Rank(A) = m, and where $A = [A_1\ A_2]$, $A_1 \in R^{m \times m}$

and is non-singular and $A_2 \varepsilon R^{m \times (n-m)}$. Then,

$$Q = \begin{pmatrix} A_1 & A_2 \\ 0 & I_{n-m} \end{pmatrix}^T \begin{pmatrix} C_{11} & C_{12} \\ C_{12}^T & C_{22} \end{pmatrix} \begin{pmatrix} A_1 & A_2 \\ 0 & I_{n-m} \end{pmatrix} \quad (3.2.1)$$

where

$C_{11} \varepsilon R^{m \times m}$ is copositive,

$C_{22} \varepsilon R^{(n-m) \times (n-m)}$ is positive semi-definite,

$C_{12}^T y \varepsilon$ Range (C_{22}) for all $y \geqslant 0$, $y \varepsilon R^m$, $C_{12} \varepsilon R^{m \times (n-m)}$

$C_{11}-C_{12}C_{22}^{+}C_{12}^T$ is copositive, where '+' denotes pseudo-inverse. (3.2.2)

Note that the rank assumption on A guarantees that there is an arrangement of variables such that A_1 is non-singular.

COROLLARY 3.2.1 A useful special case arises when $A \varepsilon R^{n \times n}$ has full rank n. In this case

$$Q - A^T CA = 0 \quad (3.2.3)$$

where $C \varepsilon R^{n \times n}$ is copositive.

COROLLARY 3.2.2 If $A \equiv 0$ then (3.2.1), (3.2.2) imply that

$$Q = C_{22} \geqslant 0 \text{ (positive semi-definite).} \quad (3.2.4)$$

There is a further corollary which is sufficiently important to warrant its statement as a theorem.

THEOREM 3.2.2 A sufficient condition for $x^TQx \geqslant 0$ for all x such that $Ax \geqslant 0$, $A \in R^{m \times n}$ is that there exists a copositive matrix $C \in R^{m \times m}$ such that

$$Q - A^TCA \geqslant 0 \text{ (positive semi-definite).} \quad (3.2.5)$$

Note that there is here no assumption on the rank of A. If A is square and non-singular, Corollary 3.2.1 shows that the condition of Theorem 3.2.2 is both sufficient and necessary.

PROOF OF THEOREM 3.2.1 We separate the vector x into two pieces as follows

$$x = \begin{pmatrix} x_1 \\ x_2 \end{pmatrix} \quad (3.2.6)$$

where $x_1 \in R^m$, $x_2 \in R^{n-m}$. We then see that

$$y \triangleq Ax = A_1x_1 + A_2x_2 \,, \quad y \in R^m, \quad y \geqslant 0 \quad (3.2.7)$$

whence

$$x_1 = A_1^{-1}(y-A_2x_2) \quad (3.2.8)$$

and it then follows that

$$x = \begin{pmatrix} A_1^{-1}(y-A_2x_2) \\ x_2 \end{pmatrix} = \begin{pmatrix} A_1^{-1} & -A_1^{-1}A_2 \\ 0 & I_{n-m} \end{pmatrix} \begin{pmatrix} y \\ x_2 \end{pmatrix}. \quad (3.2.9)$$

We then have that

$$\begin{pmatrix} y^T & x_2^T \end{pmatrix} \bar{A}^T Q \bar{A} \begin{pmatrix} y \\ x_2 \end{pmatrix} \geqslant 0 \text{ for all } y \geqslant 0, \quad x_2 \in R^{n-m} \tag{3.2.10}$$

where

$$\bar{A} = \begin{pmatrix} A_1^{-1} & -A_1^{-1}A_2 \\ 0 & I_{n-m} \end{pmatrix} . \tag{3.2.11}$$

Defining

$$\begin{pmatrix} C_{11} & C_{12} \\ C_{12}^T & C_{22} \end{pmatrix} = \bar{A}^T Q \bar{A} \tag{3.2.12}$$

we see that (3.2.10) holds if and only if conditions (3.2.2) are satisfied. Equation (3.2.1) follows upon noting that the inverse of $\bar{A}$ is

$$\begin{pmatrix} A_1 & A_2 \\ 0 & I_{n-m} \end{pmatrix} . \tag{3.2.13}$$

3.2.2 An Example

Here we illustrate Theorem 3.2.2. Let

$$Q = \begin{pmatrix} 0 & 1 \\ 1 & 0 \end{pmatrix} , \quad A = \begin{pmatrix} -1 & 1 \\ 2 & -1 \end{pmatrix} . \tag{3.2.14}$$

We wish to know whether $x^TQx \geqslant 0$ for all $x \in R^2$ such that $Ax \geqslant 0$.

Clearly Q is not positive semi-definite. However, choosing

C to be the copositive matrix

$$\begin{pmatrix} 2 & 1 \\ 1 & 0 \end{pmatrix} \tag{3.2.15}$$

yields

$$Q - A^TCA = \begin{pmatrix} 0 & 1 \\ 1 & 0 \end{pmatrix} - \begin{pmatrix} -1 & 2 \\ 1 & -1 \end{pmatrix} \begin{pmatrix} 2 & 1 \\ 1 & 0 \end{pmatrix} \begin{pmatrix} -1 & 1 \\ 2 & -1 \end{pmatrix}$$

$$= \begin{pmatrix} 0 & 1 \\ 1 & 0 \end{pmatrix} - \begin{pmatrix} -2 & 1 \\ 1 & 0 \end{pmatrix} = \begin{pmatrix} 2 & 0 \\ 0 & 0 \end{pmatrix} \tag{3.2.16}$$

which is positive semi-definite, implying the non-negativity of x^TQx for all x such that $Ax \geqslant 0$.

3.2.3 Application to Quadratic Programming

A typical quadratic programming problem is the following. Minimize the quadratic function

$$f(x) = \tfrac{1}{2}x^TQx \tag{3.2.17}$$

subject to the inequality constraints

$$Ax \geqslant b \tag{3.2.18}$$

where $x \in R^n$, $A \in R^{m \times n}$, $b \in R^m$ and $Q \in R^{n \times n}$ is symmetric.

It is well known that a necessary condition (Fritz John, Kuhn-Tucker) for $\hat{x}$ to minimize (3.2.17) subject to (3.2.18) is that there exists $\lambda \geqslant 0$, $\lambda \in R^m$, such that

$$Q\hat{x} - A^T\lambda = 0 \qquad (3.2.19)$$

and

$$A\hat{x} - b \geqslant 0 \;, \quad \lambda^T(A\hat{x}-b) = 0. \qquad (3.2.20)$$

It is also known, and easily shown, that if $\hat{x}$ satisfies (3.2.19) and (3.2.20) then $\hat{x}$ actually minimizes (3.2.17) if

$$Q \geqslant 0 \text{ (positive semi-definite).} \qquad (3.2.21)$$

The case $Q \geqslant 0$ is referred to as convex because the function $f(x)$ is convex if (3.2.21) holds. It turns out that the convex case is rather restrictive as there are interesting quadratic problems where $Q \ngeqslant 0$, and this has stimulated research on the non-convex case [7]-[13]. We are now able to provide the following useful sufficient condition for optimality in the non-convex case.

THEOREM 3.2.3. Suppose that $\hat{x}$ satisfies (3.2.19), (3.2.20). Then a sufficient condition for $\hat{x}$ to minimize (3.2.17) is that there exists a copositive matrix $C \in R^{m \times m}$ such that

$$Q - A^TCA \geqslant 0 \text{ (positive semi-definite)} \qquad (3.2.22)$$

and

$$C(A\hat{x}-b) = 0. \qquad (3.2.23)$$

PROOF If $\hat{x}$ minimizes (3.2.17) then

$$(x+\hat{x})^TQ(x+\hat{x}) \geqslant \hat{x}^TQ\hat{x} \qquad (3.2.24)$$

for all x such that

$$A(\hat{x}+x) - b \geqslant 0. \qquad (3.2.25)$$

From (3.2.22), (3.2.23) we have

$$x^TQx - [A(\hat{x}+x)-b]^TC[A(\hat{x}+x)-b] \geqslant 0, \text{ for all } x \in R^n \qquad (3.2.26)$$

and from (3.2.19), (3.2.20) that

$$x^TQ\hat{x} = \lambda^T[A(\hat{x}+x)-b] \text{ , for all } x \in R^n. \qquad (3.2.27)$$

From (3.2.26) and (3.2.27) we see that both x^TQx and $x^TQ\hat{x}$ are non-negative for all x such that (3.2.25) holds. Inequality (3.2.24) now follows easily.

Note that if we further restrict C to be a matrix with all its elements non-negative, equation (3.2.23) becomes a complementary slackness condition similar to the second inequality of (3.2.20).

3.2.4 Stochastically Copositive Matrices

Theorem 3.1.1 is remarkable in that it characterizes copositive matrices only for $n \leqslant 4$. In higher dimensions such matrices remain uncharacterized although certain extreme forms are known [4], [5].

In this section we demonstrate that stochastically copositive matrices are of the form S + P for all n.

DEFINITION 3.2.1 A symmetric matrix $Q \in R^{n \times n}$ is stochastically copositive if

$$E[x^TQx] \geqslant 0 \qquad (3.2.28)$$

for all covariance matrices ($X \geqslant 0$) such that

$$X_{ij} \geqslant 0 \ , \quad i,j=1,\ldots,n \tag{3.2.29}$$

where $x \in R^n$ is a random variable with zero mean and covariance matrix X.

We have the following theorem.

THEOREM 3.2.4 $Q \in R^{n \times n}$ is stochastically copositive if and only if Q can be decomposed as

$$Q = S + P \tag{3.2.30}$$

where $S \in R^{n \times n}$ is positive semi-definite and $P \in R^{n \times n}$ has non-negative elements.

COROLLARY 3.2.3 If $Q \in R^{n \times n}$ is stochastically copositive then it is copositive.

COROLLARY 3.2.4 Every copositive $Q \in R^{n \times n}$ is stochastically copositive if and only if $n \leqslant 4$.

It follows that copositivity and stochastic copositivity are equivalent if and only if $n \leqslant 4$.

PROOF OF THEOREM 3.2.4 It is easy to see that (3.2.28) is equivalent to

$$\mathrm{tr}(QX) \geqslant 0 \tag{3.2.31}$$

for all $X \geqslant 0$, $X_{ij} \geqslant 0$, $i,j=1,\ldots,n$.

The fact that (3.2.31) follows from (3.2.30) is trivial. We now prove the converse. Define the following sets in $R \times R^{n \times n}$

$$\tilde{A} \triangleq \{r,Z \mid r \geq tr(QX),\ Z_{ij} \leq X_{ij},\ i,j=1,\ldots,n,\ \text{for some } X \geq 0\} \tag{3.2.32}$$

$$\tilde{B} \triangleq \{r,Z \mid r \leq 0,\ Z_{ij} \geq 0,\ i,j=1,\ldots,n\}. \tag{3.2.33}$$

Clearly the sets $\tilde{A}$ and $\tilde{B}$ are both convex. $\tilde{A}$ does not contain interior points of $\tilde{B}$. For, supposing the contrary, there would exist $X \geq 0$, $X_{ij} > 0$, $i,j=1,\ldots,n$ such that $tr(QX) < 0$, contradicting (3.2.31). Thus there exists a hyperplane with parameters ρ and $-P$ which separates $\tilde{A}$ and $\tilde{B}$. As $(r=0,\ Z=0)$ belongs to both $\tilde{A}$ and $\tilde{B}$ we have

$$\begin{aligned} \rho r - tr(PZ) &\geq 0,\ (r,Z) \in \tilde{A} \\ &\leq 0,\ (r,Z) \in \tilde{B}. \end{aligned} \tag{3.2.34}$$

By the nature of $\tilde{B}$ it follows that $\rho \geq 0$ and $P_{ij} \geq 0$, $i,j=1,\ldots,n$ and since there exists $X \geq 0$, $X_{ij} > 0$, $i,j=1,\ldots,n$ we see that ρ can be set equal to unity. Thus

$$tr(QX) - tr(PX) \geq 0, \text{ for all } X \geq 0 \tag{3.2.35}$$

and the theorem is proved.

3.2.5 Symmetric M-Matrices

Theorem 3.1.2, the dual of Theorem 3.1.1, states that all positive semi-definite matrices with non-negative elements are completely positive if and only if $n \leq 4$. In fact there are known determinantal conditions [6] which hold for all n.

In this section we investigate the factorization of symmetric

M-matrices. These matrices are positive definite and their inverses are matrices with non-negative elements. We shall show that these inverses are in fact completely positive for all n.

DEFINITION 3.2.2 An M-matrix $Q \in R^{n \times n}$ is such that $q_{ij} \leqslant 0$, $i \neq j$, together with one of the following three equivalent properties:

$$\text{Q is non-singular and the elements of } Q^{-1} \text{ are positive} \tag{3.2.36}$$

$$\text{all principal minors of Q are positive} \tag{3.2.37}$$

$$\text{there exists a vector } x > 0,\ x \in R^n, \text{ such that } Qx > 0. \tag{3.2.38}$$

THEOREM 3.2.5 [14] Every symmetric M-matrix $Q \in R^{n \times n}$ has the representation $Q = GG^T$, where G is a triangular M-matrix.

PROOF The proof is by induction. Let Q_n be an n-dimensional symmetric M-matrix. As it is symmetric it is also positive definite, (3.2.37). We now suppose that the theorem is true for (n-1). We write

$$Q_n = \begin{pmatrix} Q_{n-1} & b \\ b^T & q_{nn} \end{pmatrix} \tag{3.2.39}$$

where $b \leqslant 0$ and Q_{n-1} is a symmetric M-matrix of dimension n-1. By hypothesis there is a triangular M-matrix G_{n-1} such that

$$G_{n-1}G_{n-1}^T = Q_{n-1}. \tag{3.2.40}$$

As G_{n-1} is non-singular and $b \leqslant 0$ there exists $c \leqslant 0$ such that

$$G_{n-1}c = b. \tag{3.2.41}$$

Now for any scalar x we have

$$\begin{pmatrix} G_{n-1} & 0 \\ c^T & x \end{pmatrix} \begin{pmatrix} G_{n-1}^T & c \\ 0 & x \end{pmatrix} = \begin{pmatrix} Q_{n-1} & b \\ b^T & c^Tc+x^2 \end{pmatrix}. \tag{3.2.42}$$

If it is possible to define $x > 0$ by the equation

$$c^Tc + x^2 = q_{nn} \tag{3.2.43}$$

we have a triangular M-matrix decomposition of Q_n. Indeed, from the fact that

$$\begin{aligned} q_{nn} - c^Tc &= q_{nn} - b^T(G_{n-1}G_{n-1}^T)^{-1}b \\ &= q_{nn} - b^TQ_{n-1}^{-1}b = \det(Q_n)/\det(Q_{n-1}) > 0 \end{aligned} \tag{3.2.44}$$

we see that the definition of x via (3.2.43) is valid. The theorem follows by induction since it is obviously true for $n = 2$.

THEOREM 3.2.6 The inverse of every symmetric M-matrix $Q \in R^{n \times n}$ is completely positive. Moreover, it has the factorization $Q^{-1} = FF^T$ with F a triangular matrix having non-negative elements.

PROOF The proof follows immediately upon noting that G^{-1} has non-negative elements and is triangular.

3.2.6 Generalization of Finsler's Theorem

In 1937 P. Finsler [15] proved that if $x^TQx > 0$ for all x such that $x^TAx = 0$, where Q, A are symmetric real matrices, then there exists a real scalar σ such that

$$Q - \sigma A > 0 \text{ (positive definite).} \qquad (3.2.45)$$

In this section we generalize Finsler's theorem by treating the more general case of an arbitrary number of quadratic equalities and inequalities. Furthermore we apply the generalized result to deduce certain properties of the inverse of a copositive matrix.

We consider the following formulation

$$\text{Minimize } x^TQx \qquad (3.2.46)$$

subject to

$$x^TA_ix = b_i\ , \quad i=1,\ldots,r \qquad (3.2.47)$$

and

$$x^TC_jx \geqslant d_j\ , \quad j=1,\ldots,s \qquad (3.2.48)$$

where Q, A_i, $i=1,\ldots,r$ and C_j, $j=1,\ldots,s$ are $n \times n$ symmetric matrices and $x \in R^n$. We assume that there is an $x \in R^n$ which satisfies (3.2.47), (3.2.48).

THEOREM 3.2.7 [16] Suppose that there exists a non-singular matrix F which simultaneously diagonalizes Q, A_i, i=1,...,r, C_j, j=1,...,s. Then the non-linear programming problem formulated above has a solution $\hat{x} \in R^n$ if and only if there exist scalars σ_i, i=1,...,r, $\rho_j \geq 0$, j=1,...,s and a diagonal matrix $\bar{N}$ having non-negative elements, such that

$$Q - \sum_{i=1}^{r} \sigma_i A_i - \sum_{j=1}^{s} \rho_j C_j - (F^{-1})^T \bar{N} F^{-1} = 0 \qquad (3.2.49)$$

$$\sum_{j=1}^{s} \rho_j [\hat{x}^T C_j \hat{x} - d_j] = 0 \qquad (3.2.50)$$

$$\hat{x}^T (F^{-1})^T \bar{N} F^{-1} \hat{x} = 0 \qquad (3.2.51)$$

$$\hat{x}^T A_i \hat{x} = b_i \ , \quad i=1,\dots,r \qquad (3.2.52)$$

$$\hat{x}^T C_j \hat{x} \geq d_j \ , \quad j=1,\dots,s. \qquad (3.2.53)$$

COROLLARY 3.2.5 Under the stated assumption of simultaneous diagonalizability we have that $x^T Q x \geq 0$ for all $x \in R^n$ such that $x^T A_i x = 0$, i=1,...,r, $x^T C_j x \geq 0$, j=1,...,s if and only if there exist real scalars σ_i, i=1,...,r and $\rho_j \geq 0$, j=1,...,s such that

$$Q - \sum_{i=1}^{r} \sigma_i A_i - \sum_{j=1}^{s} \rho_j C_j \geq 0 \text{ (positive semi-definite)} . \qquad (3.2.54)$$

PROOF OF THEOREM 3.2.7 Let $y = F^{-1}x$. Then, because $\bar{Q} \triangleq F^T Q F$, $\bar{A}_i \triangleq F^T A_i F$, i=1,...,r, $\bar{C}_j \triangleq F^T C_j F$, j=1,...,s are diagonal, the above formulation becomes

$$\underset{y_1^2,\ldots,y_n^2}{\text{Minimize}} \sum_{k=1}^{n} \bar{Q}_{kk} y_k^2 \tag{3.2.55}$$

subject to

$$\sum_{k=1}^{n} (\bar{A}_i)_{kk} y_k^2 = b_i \ , \quad i=1,\ldots,r \tag{3.2.56}$$

$$\sum_{k=1}^{n} (\bar{C}_j)_{kk} y_k^2 \geqslant d_j \ , \quad j=1,\ldots,s. \tag{3.2.57}$$

Re-naming $y_1^2,\ldots,y_n^2$ as $z_1 \geqslant 0,\ldots,z_n \geqslant 0$ we see that the problem is

$$\underset{z_1,\ldots,z_k}{\text{Minimize}} \sum_{k=1}^{n} \bar{Q}_{kk} z_k \tag{3.2.58}$$

subject to

$$\sum_{k=1}^{n} (\bar{A}_i)_{kk} z_k = b_i \ , \quad i=1,\ldots,r \tag{3.2.59}$$

$$\sum_{k=1}^{n} (\bar{C}_j)_{kk} z_k \geqslant d_j \ , \quad j=1,\ldots,s \tag{3.2.60}$$

$$z_k \geqslant 0 \ , \quad k=1,\ldots,n \ . \tag{3.2.61}$$

Well-known necessary and sufficient conditions for $\hat{z}_1,\ldots,\hat{z}_n$ to solve this linear program are that there exist σ_i, $i=1,\ldots,r$ and $\rho_j \geqslant 0$, $j=1,\ldots,s$ such that

$$\bar{Q} - \sum_{i=1}^{r} \sigma_i \bar{A}_i - \sum_{j=1}^{s} \rho_j \bar{C}_j - \bar{N} = 0 \tag{3.2.62}$$

where the diagonal matrix $\bar{N}$ has non-negative elements and where

$$\sum_{j=1}^{s} \rho_j [\sum_{k=1}^{n} (\bar{C}_j)_{kk}\hat{z}_k - d_j] = 0 \qquad (3.2.63)$$

$$\sum_{k=o}^{n} \bar{N}_{kk}\hat{z}_k = 0. \qquad (3.2.64)$$

Pre-multiplying (3.2.62) by (F^{-1}) and post-multiplying by F^{-1} yields (3.2.49), and (3.2.63) and (3.2.64) become (3.2.50), (3.2.51) respectively.

We now establish the relationship of the above results to that of Finsler (3.2.45). First we note the important fact that any matrices Q and A which satisfy Finsler's conditions also satisfy our assumption of simultaneous diagonalizability. This follows from the fact that Q and $(Q-\sigma A)$, which is positive definite, are simultaneously diagonalizable by a non-singular matrix [17]. Next we note that (3.2.45) is recoverable from Corollary 3.2.5. For setting $A_1 = A$, $A_2 = \ldots = A_r = 0$, $C_1 = \ldots = C_s = 0$ we have that there is a σ_1 such that

$$Q - \sigma_1 A \geqslant 0. \qquad (3.2.65)$$

It is easy to verify that the inequality in (3.2.65) can be strengthened to strict inequality.

From the above discussion it is clear that Theorem 3.2.7 and Corollary 3.2.5 are non-trivial generalizations of Finsler's Theorem.

We now investigate the degree of restrictiveness of the

assumption of simultaneous diagonalizability. The notion of copositiveness assists us here.

THEOREM 3.2.8 Theorem 3.2.7 does not necessarily hold if the assumption of simultaneous diagonalizability is removed.

PROOF The proof is by a counter-example. Suppose that $Q \in R^{5 \times 5}$ is copositive. Then the statement that $x^TQx \geqslant 0$ for all $x \geqslant 0$ can be re-phrased as

$$x^TQx \geqslant 0 \tag{3.2.66}$$

subject to

$$x_ix_j \geqslant 0, \quad i=1,\ldots,4, \quad j=i+1,\ldots,5. \tag{3.2.67}$$

If Theorem 3.2.7 were here valid we would have

$$Q - \sum_{j=1}^{10} \rho_jC_j \geqslant 0 \tag{3.2.68}$$

where the C_j, $j=1,\ldots,10$ are the matrices with non-negative elements which realize (3.2.67). However, (3.2.68) would then imply that every copositive matrix in five dimensions is expressible as the sum of a positive semi-definite matrix and a matrix with non-negative elements. But the example of (3.1.6) disproves this and so we conclude that Theorem 3.2.7 is not necessarily valid in the absence of the stated assumption.

The next theorem provides necessary conditions and sufficient conditions for (3.2.54) to hold; the gap between necessity and sufficiency is small.

THEOREM 3.2.9 (i) Necessary Condition. Suppose that (3.2.54) holds with strict inequality (i.e. it is positive-definite). The Q and $(\sum_{i=1}^{r} \sigma_i A_i + \sum_{j=1}^{s} \rho_j C_j)$ are simultaneously diagonalizable and $x^T Q x \geqslant 0$ for all $x \in R^n$ such that $x^T(\sum_{i=1}^{r} \sigma_i A_i + \sum_{j=1}^{s} \rho_j C_j)x \geqslant 0$.

(ii) Sufficient Condition. Suppose there exist $\hat{\sigma}_i$, i=1,...,r and $\hat{\rho}_j$, j=1,...,s such that Q and $(\sum_{i=1}^{r} \hat{\sigma}_i A_i + \sum_{j=1}^{s} \hat{\rho}_j C_j)$ are simultaneously diagonalizable and that $x^T Q x \geqslant 0$ for all $x \in R^n$ such that $x^T(\sum_{i=1}^{r} \hat{\sigma}_i A_i + \sum_{j=1}^{s} \hat{\rho}_j C_j)x \geqslant 0$. Then inequality (3.2.54) holds.

The above theorem, which is easy to prove, indicates the extent to which the diagonalizability assumption of Corollary 3.2.5 may be weakened.

3.2.7 Application of Generalized Finsler Theorem

We establish here certain properties of inverses of copositive and other matrices. First we present necessary and sufficient conditions for

$$x^T Q^{-1} x \geqslant 0 \tag{3.2.69}$$

for all $x \in R^n$ such that

$$x^T Q x \geqslant 0. \tag{3.2.70}$$

THEOREM 3.2.10 A necessary and sufficient condition for (3.2.69) to hold for all x such that (3.2.70) holds is that there is a scalar $\rho \geqslant 0$ such that

$$Q - \rho Q^3 \geqslant 0. \tag{3.2.71}$$

Written in terms of the eigenvalues λ_k, k=1,...,n of Q this is

$$\lambda_k - \rho\lambda_k^3 \geqslant 0. \tag{3.2.72}$$

COROLLARY 3.2.6 A sufficient condition for the inverse of a copositive matrix Q to be copositive is that (3.2.71) holds for some $\rho \geqslant 0$.

PROOF OF THEOREM 3.2.10 The problem as posed is clearly equivalent to that of the non-negativity of x^TQx for all $x \in R^n$ such that $x^TQ^3x \geqslant 0$. Because of the fact that Q and Q^3 commute, there exists an orthogonal (invertible) matrix which simultaneously diagonalizes them. Corollary 3.2.5 therefore applies, yielding (3.2.71).

A further interesting observation is that Theorem 3.2.10 implies that necessary and sufficient conditions for the set $\{x \in R^n | x^TQx \geqslant 0\}$ to be identical to the set $\{x \in R^n | x^TQ^{-1}x \geqslant 0\}$ are that there exist $\rho_1 \geqslant 0$, $\rho_2 \geqslant 0$ such that

$$Q - \rho_1 Q^3 \geqslant 0$$

and

$$Q^3 - \rho_2 Q \geqslant 0. \tag{3.2.73}$$

These inequalities imply that $Q(1-\rho_1\rho_2) \geqslant 0$ which implies that Q is positive definite or negative definite, or that $\rho_1\rho_2 = 1$. This last-mentioned case implies that

$$Q - \rho_1 Q^3 \geqslant 0, \quad \rho_1 Q^3 - Q \geqslant 0 \tag{3.2.74}$$

or that

$$Q^{-1} = \rho_1 Q, \; \rho_1 > 0. \tag{3.2.75}$$

An example of such a matrix in R^2 is $\begin{pmatrix} 0 & 2 \\ 2 & 0 \end{pmatrix}$.

3.3 Quadratic Differential Equations

In this section we investigate the properties of a general class of quadratic differential equations. More specifically, we obtain sufficient conditions for the existence of finite escape times and provide upper bounds for these times. The class of problems studied is of interest in its own right but appears to be of importance also in optimal control theory (cf. Section 2) and in the mathematical modelling of certain chemical engineering and predator-prey problems.

It turns out that there exist certain Liapunov-type theorems which provide sufficient conditions for the existence of finite escape times in solutions of non-linear differential equations [18] but these theorems have not directly yielded useful results in the case of quadratic differential equations. The primary reason for this is that the time derivative of a scalar function of even order is odd, and therefore it is difficult to comment upon its sign without having some further knowledge of the trajectories of the quadratic differential equations. Furthermore the known general theorems mostly assume positivity and radial unboundedness of a Liapunov-like function, which are properties usually associated with even functions. Thus few explicit results are known for quadratic differential equations.

Our approach to the problem is to use a 'linear Liapunov function' whose time derivative is then a quadratic function of the system state. Requiring this quadratic form to have a definite sign for all values of the system state or, less restrictively, for those values of the state which lie in certain invariant sets (cones) allows us to obtain our sufficient conditions. We apply our conditions to the well-known matrix Riccati equation to obtain results which complement those of Kalman. We demonstrate also that the conditions can be applied to a general class of quadratic equations used to model interacting populations. The Lotka-Volterra model of predator-prey dynamics is a special case of this class. Note that the idea of using a linear Liapunov function and requiring that its time derivative be definite in sign for all values of the state has occurred also to Frayman [19], but our first results pre-date his [20].

3.3.1 Finite Escape Times

Consider the general non-linear system of ordinary differential equations

$$\dot{x} = f(x,t), \quad x(0) = x_0 \tag{3.3.1}$$

where $x \in R^n$ and where $f:R^{n+1} \to R^n$ is sufficiently well-behaved to guarantee a unique solution $x(t;x_0)$ on $[0,\hat{t})$ for some $\hat{t} > 0$. It is known that either $x(t;x_0)$ may be extended for all t beyond $\hat{t}$ or there exists a time $t_e < \infty$ such that $\|x(t;x_0)\| \to \infty$ as $t \to t_e$. In the latter case the solution is said to have a finite escape time at t_e.

In this chapter we derive sufficient conditions for the solution of the quadratic equations

$$\dot{x}_i = x^T A_i x + b_i^T x + c_i, \quad i=1,\ldots,n, \quad x(0) = x_0 \tag{3.3.2}$$

to have a finite escape time. Here, the matrices $A_i \in R^{n \times n}$, $i=1,\ldots,n$ are constant, and with no loss of generality are assumed to be symmetric. The vectors $b_i \in R^n$ and the scalars c_i, $i=1,\ldots,n$ are constant.

Owing to the fact that the right-hand sides of the differential equations (3.3.2) are quadratic in x, it follows from standard existence theory that for each $x_0 \in R^n$ there exists a unique solution $x(t;x_0)$ of (3.3.2) defined on the interval $[0,\hat{t}(x_0))$ where $\hat{t}(x_0) > 0$ for each $x_0 \in R^n$. Furthermore it is known that even the simplest quadratic differential equation can exhibit finite escape times. For example the scalar quadratic differential equation

$$\dot{y} = y^2 \tag{3.3.3}$$

has a unique solution which has a finite escape time $t_e = \frac{1}{x_0}$ if and only if $x_0 > 0$.

Before proceeding to an analysis of (3.3.2) we consider in detail the scalar quadratic equation. The results so obtained are then used to bound the solutions of (3.3.2).

LEMMA 3.3.1 Given constants $k_1 > 0$, k_2, k_3 the solution $y(t;y_0)$ of the scalar differential equation

$$\dot{y} = k_1 y^2 + k_2 y + k_3, \quad y(0) = y_0 \tag{3.3.4}$$

goes to infinity in a finite time under the following conditions:

(i) for all y_0, whenever $k_2^2 - 4k_1k_3 < 0$

(ii) for $y_0 > (-k_2 + \sqrt{k_2^2-4k_1k_3})/2k_1$, whenever $k_2^2 - 4k_1k_3 \geq 0$.

Let α, β with $\alpha \geq \beta$ be the real roots of $k_1y^2 + k_2y + k_3 = 0$ whenever $\Delta \triangleq k_2^2 - 4k_1k_3 \geq 0$. Then the escape time t_e in case (i) is given by

$$t_e = \frac{2}{\sqrt{-\Delta}} \left\{\frac{\pi}{2} - \arctan[2k_1(y_0 + \frac{k_2}{2k_1})/\sqrt{-\Delta}]\right\} \tag{3.3.5}$$

and in case (ii) is given by

$$t_e = \frac{1}{\sqrt{\Delta}} \cdot \ln[(y_0-\beta)/(y_0-\alpha)], \quad \Delta \neq 0 \tag{3.3.6}$$

and

$$t_e = 2/(2k_1y_0+k_2), \quad \Delta = 0. \tag{3.3.7}$$

PROOF Local existence theory guarantees that there is a $\hat{t} > 0$ such that

$$t = \int_{y_0}^{y(t)} \frac{dy}{k_1y^2+k_2y+k_3}, \quad t \in [0,\hat{t}). \tag{3.3.8}$$

This expression can be integrated to yield $y(t)$, but there are two distinct cases.

(i) If $\Delta < 0$ we have for $t \varepsilon [0,\hat{t})$,

$$y(t;y_0) = \frac{\sqrt{-\Delta}}{2k_1} \tan\left[\frac{t\sqrt{-\Delta}}{2} + \arctan\left(\frac{2k_1 y_0 + k_2}{\sqrt{-\Delta}}\right)\right] - \frac{k_2}{2k_1} \, . \tag{3.3.9}$$

Without loss of generality we consider the principal value of arctan $(\cdot)$ so that

$$-\frac{\pi}{2} < \arctan\left(\frac{2k_1 y_0 + k_2}{\sqrt{-\Delta}}\right) < \frac{\pi}{2} \, . \tag{3.3.10}$$

Also since $k_1 > 0$ we have $y(t) \to \infty$ as $t \to t_e$ where t_e satisfies

$$\frac{t_e\sqrt{-\Delta}}{2} + \arctan\left(\frac{2k_1 y_0 + k_2}{\sqrt{-\Delta}}\right) = \frac{\pi}{2} \, . \tag{3.3.11}$$

Clearly the value of t_e depends upon y_0, but the existence of t_e does not. Indeed, as $y_0 \to -\infty, \arctan\left(\frac{2k_1 y_0 + k_2}{\sqrt{-\Delta}}\right) \to -\frac{\pi}{2}$ and the maximum escape time for any y_0 is therefore less than $\frac{2\pi}{\sqrt{-\Delta}}$. Solving (3.3.11) for t_e yields (3.3.5).

(ii) If $\Delta > 0$ we have

$$y(t;y_0) = \frac{\alpha(y_0-\beta) - \beta(y_0-\alpha)e^{\sqrt{\Delta}\,t}}{(y_0-\beta) - (y_0-\alpha)e^{\sqrt{\Delta}\,t}} \, , \quad t \varepsilon [0,\hat{t}). \tag{3.3.12}$$

Note that if $(y_0-\alpha) < 0$, then $y(t;y_0)$ can be extended for all $t \in [0,\infty)$. However, if $(y_0-\alpha) > 0$ the solution can only be extended as far as t_e where t_e satisfies

$$(y_0-\beta) - (y_0-\alpha)e^{\sqrt{\Delta}\, t_e} = 0 \qquad (3.3.13)$$

which yields (3.3.6). Equation (3.3.7) is obtained upon direct integration of (3.3.8).

The next lemma relates a certain differential inequality to (3.3.4).

LEMMA 3.3.2 Suppose that there exists a scalar function $V(x)$ of $x \in R^n$ and scalars $k_1 > 0$, k_2 and k_3 such that along solutions of (3.3.1)

$$\frac{dV(x)}{dt} \geqslant k_1V^2(x) + k_2V(x) + k_3 \,. \qquad (3.3.14)$$

Suppose further that $\|x\| \to \infty$ whenever $V(x) \to \infty$. Then the solution $x(t;x_0)$ of (3.3.1) has a finite escape time whenever the solution of

$$\dot{y} = k_1y^2 + k_2y + k_3, \quad y(0) = V(x_0) \qquad (3.3.15)$$

goes to infinity in a finite time. Furthermore Lemma 3.3.1 provides upper bounds for the escape time of (3.3.1).

PROOF It follows immediately from (3.3.14) and (3.3.15) that

$$V(x(t;x_0)) \geqslant y(t;V(x_0)) \qquad (3.3.16)$$

so that V goes to infinity in a finite time if y does, and this implies that $\|x(t;x_0)\| \to \infty$.

We may now use Lemmas 3.3.1 and 3.3.2 to deduce sufficient conditions for the solution of (3.3.2) to have a finite escape time. For any $\omega \in R^n$ with i-th element ω_i, we define

$$A_\omega \triangleq \sum_{i=1}^{n} \omega_i A_i \tag{3.3.17}$$

$$b_\omega \triangleq \sum_{i=1}^{n} \omega_i b_i \tag{3.3.18}$$

$$c_\omega \triangleq \sum_{i=1}^{n} \omega_i c_i \tag{3.3.19}$$

$$\Delta_\omega = (\lambda_\omega/\omega^T\omega)(b_\omega^T A_\omega^{-1} b_\omega - 4c_\omega) \tag{3.3.20}$$

where λ_ω is the smallest eigenvalue of the symmetric matrix A_ω.

THEOREM 3.3.1 [21] Suppose there exists a vector $\omega \in R^n$ such that $A_\omega > 0$ (positive definite). Then the solution of (3.3.2) has a finite escape time t_e under the following conditions

(i) for all x_0, whenever $\Delta_\omega < 0$;

(ii) for x_0 satisfying

$$\omega^T x_0 > -\tfrac{1}{2}\omega^T A_\omega^{-1} b_\omega + \tfrac{1}{2}\sqrt{\Delta_\omega}\cdot \frac{\omega^T\omega}{\lambda_\omega} \tag{3.3.21}$$

whenever $\Delta_\omega \geqslant 0$.

In case (i)

$$t_e \leqslant \frac{2}{\sqrt{-\Delta_\omega}} \{\frac{\pi}{2} - \arctan[2\omega^T(x_0+\tfrac{1}{2}A_\omega^{-1}b_\omega)\frac{\lambda_\omega}{\omega^T\omega} \cdot \frac{1}{\sqrt{-\Delta_\omega}}]\}. \tag{3.3.22}$$

In case (ii)

$$t_e \leqslant \frac{1}{\sqrt{\Delta_\omega}} \ln \left(\frac{\omega^T(x_0+\frac{1}{2}A_\omega^{-1}b_\omega)+\frac{1}{2}\sqrt{\Delta_\omega}\cdot\omega^T\omega/\lambda_\omega}{\omega^T(x_0+\frac{1}{2}A_\omega^{-1}b_\omega)-\frac{1}{2}\sqrt{\Delta_\omega}\cdot\omega^T\omega/\lambda_\omega}\right). \tag{3.3.23}$$

PROOF We obtain from Equation (3.3.2) that

$$\frac{d}{dt}(\omega^T x) = x^T A_\omega x + b_\omega^T x + c_\omega \tag{3.3.24}$$

which may be written as

$$\begin{aligned}\frac{d}{dt}(\omega^T x) &= (x+\tfrac{1}{2}A_\omega^{-1}b_\omega)^T A_\omega (x+\tfrac{1}{2}A_\omega^{-1}b_\omega) + c_\omega - \tfrac{1}{4}b_\omega^T A_\omega^{-1} b_\omega \\ &\geqslant \frac{\lambda_\omega}{\omega^T\omega} \cdot [\omega^T(x+\tfrac{1}{2}A_\omega^{-1}b_\omega)]^2 + c_\omega - \tfrac{1}{4}b_\omega^T A_\omega^{-1} b_\omega\end{aligned} \tag{3.3.25}$$

by the Cauchy-Schwartz inequality.

Re-arranging (3.3.25) we obtain

$$\begin{aligned}\frac{d}{dt}(\omega^T x) \geqslant \frac{\lambda_\omega}{\omega^T\omega} \cdot (\omega^T x)^2 + \frac{\lambda_\omega}{\omega^T\omega} \cdot \omega^T A_\omega^{-1} b_\omega (\omega^T x) + \frac{\lambda_\omega}{\omega^T\omega} \cdot (\tfrac{1}{2}\omega^T A_\omega^{-1} b_\omega)^2 \\ + c_\omega - \tfrac{1}{4}b_\omega^T A_\omega^{-1} b_\omega .\end{aligned} \tag{3.3.26}$$

Hence the solution of (3.3.2) has a finite escape time under the conditions provided in Lemma 3.3.1 if we identify k_1, k_2 and k_3 as follows

$$k_1 \triangleq \frac{\lambda_\omega}{\omega^T\omega} > 0 \qquad (3.3.27)$$

$$k_2 \triangleq \frac{\lambda_\omega}{\omega^T\omega} \cdot \omega^T A_\omega^{-1} b_\omega \qquad (3.3.28)$$

$$k_3 \triangleq \frac{\lambda_\omega}{\omega^T\omega} \cdot (\tfrac{1}{2}\omega^T A_\omega^{-1} b_\omega)^2 + c_\omega - \tfrac{1}{4} b_\omega^T A_\omega^{-1} b_\omega \qquad (3.3.29)$$

whence

$$\Delta = k_2^2 - 4k_1k_3 = \frac{\lambda_\omega}{\omega^T\omega} \cdot (b_\omega^T A_\omega^{-1} b_\omega - 4c_\omega) = \Delta_\omega. \qquad (3.3.30)$$

3.3.2 Matrix Riccati Equation

We now apply Theorem 3.3.1 to the autonomous matrix Riccati differential equation

$$\dot{P} = Q + PF + F^T P - PRP, \quad P(0) = \Pi \qquad (3.3.31)$$

where all the matrices are in $R^{n \times n}$, and Q, R, Π and hence $P(t;\Pi)$ are symmetric.

Many sufficient conditions are known for the solution of (3.3.31) to exist on $[0,\infty)$, but those known best are due to Kalman [22], viz.

$$R \geqslant 0, \quad Q \geqslant 0, \quad \Pi \geqslant 0. \qquad (3.3.32)$$

The sufficiency results for the solution of (3.3.31) to escape in a finite time, presented below, complement the existence conditions.

In order to convert (3.3.31) to a suitable vector quadratic

equation we use the following matrix operations.

(i) The Kronecker product $A \odot B$ of two matrices $A \in R^{m \times n}$ and $B \in R^{p \times q}$ is defined [17] as a matrix $C \in R^{mp \times nq}$ whose elements in the $(i-1)p+1$ to ip rows and $(j-1)q+1$ to jq columns are $(a_{ij}B)$, $i=1,\ldots,m$, $j=1,\ldots,n$, viz.

$$C = \begin{pmatrix} a_{11}B & \cdots & a_{1n}B \\ \vdots & & \vdots \\ a_{m1}B & \cdots & a_{mn}B \end{pmatrix} . \tag{3.3.33}$$

It is easily verified that

$$(A \odot B)^T = A^T \odot B^T \tag{3.3.34}$$

and

$$(A \odot B)^{-1} = A^{-1} \odot B^{-1}, \quad A, B \text{ non-singular.} \tag{3.3.35}$$

(ii) For any matrix $A \in R^{m \times n}$ define a stacking operator $\sigma(A)$ as an mn vector formed by stacking the columns of A under each other, in order, as follows

$$\sigma(A) \triangleq (a_{11},\ldots,a_{m1};a_{12},\ldots,a_{m2};\ldots;a_{1n},\ldots,a_{mn})^T . \tag{3.3.36}$$

Clearly, σ is a linear operator from $R^{m \times n} \to R^{mn}$, i.e. for $A, B \in R^{m \times n}$ and $c \in R^1$

$$\sigma(A+B) = \sigma(A) + \sigma(B) \tag{3.3.37}$$

$$\sigma(cB) = c\sigma(B). \tag{3.3.38}$$

(iii) For any matrices A, B and P for which the matrix product APB is defined, it can be shown that

$$\sigma(APB) = (B^T \bigcirc A)\sigma(P). \tag{3.3.39}$$

We now use the above operations and the symmetry of P, Q, R and Π to prove the following lemma.

LEMMA 3.3.3 Let $W \varepsilon R^{n \times n}$ be a given constant matrix. Then along solutions of (3.3.31) we have that

$$\frac{d}{dt}(\omega^T p) = p^T R_\omega p + f_\omega^T p + \omega^T q \tag{3.3.40}$$

where

$$\left.\begin{aligned} \omega &= \sigma(W) \\ p &= \sigma(P) \\ q &= \sigma(Q) \\ f_\omega &= [F \bigcirc I + I \bigcirc F]\sigma(W) \\ R_\omega &= - W \bigcirc R \,. \end{aligned}\right\} \tag{3.3.41}$$

PROOF We have that

$$\frac{d}{dt}[\sigma^T(W)\sigma(P)] = \sigma^T(W)\sigma(Q+PF+F^TP-PRP) \,. \tag{3.3.42}$$

Now

$$\begin{aligned} \sigma^T(W)\sigma(PRP) &= \sigma^T(W)(I \bigcirc PR)\sigma(P) \\ &= \sigma^T(RPW)\sigma(P) \\ &= \sigma^T(P)(W \bigcirc R)\sigma(P) \,. \end{aligned} \tag{3.3.43}$$

Similarly

$$\sigma(PF+F^TP) = (F^T \circ I + I \circ F^T)\sigma(P). \qquad (3.3.44)$$

Using these identities in (3.3.42) yields (3.3.40).

We can now apply Theorem 3.3.1 to yield

THEOREM 3.3.2 [21] The solution of (3.3.31) has a finite escape time if there exists a symmetric matrix $W \in R^{n \times n}$ such that $- W \circ R > 0$ and

(i) $\Delta(W,F,R,Q) < 0$

or

(ii) $\Delta(W,F,R,Q) \geqslant 0$ and

$$\sigma^T(W)\sigma(\Pi) > \tfrac{1}{2}\sigma^T(W)(W^{-1} \circ R^{-1})(F \circ I + I \circ F)\sigma(W)$$

$$+ \tfrac{1}{2} \sqrt{\Delta(W,F,R,Q)}\,\sigma^T(W)\sigma(W)/\lambda \qquad (3.3.45)$$

where

$$\Delta(W,F,R,Q) =$$

$$- \frac{\lambda}{\sigma^T(W)\sigma(W)} \cdot [\sigma^T(W)(F \circ I + I \circ F)^T(W^{-1} \circ R^{-1})(F \circ I + I \circ F)\sigma(W)$$

$$+ 4\sigma^T(W)\sigma(Q)] \qquad (3.3.46)$$

and where $\lambda > 0$ is the smallest eigenvalue of $- W \circ R$.

3.3.3 An Example

We apply Theorem 3.3.2 to an illustrative example in $R^{2 \times 2}$. Let $P \in R^{2 \times 2}$, $R = 2F = I_2$, and

$$Q = \begin{pmatrix} q_{11} & q_{12} \\ q_{12} & q_{22} \end{pmatrix}, \quad \Pi = \begin{pmatrix} \Pi_{11} & \Pi_{12} \\ \Pi_{12} & \Pi_{22} \end{pmatrix} \tag{3.3.47}$$

and let I_n denote the n x n identity matrix. With the above values of R and F we have

$$\dot{P} = -P^2 + P + Q, \quad P(0) = \Pi . \tag{3.3.48}$$

The Kalman conditions $\Pi \geqslant 0$, $Q \geqslant 0$ would guarantee the existence of P for all $t \in [0,\infty)$. However, let us assume that these conditions do not hold, and let us apply Theorem 3.3.2.

We chose $W = -I_2$ so that

$$-W \bigcirc R = I_4 \Rightarrow \lambda = 1 \tag{3.3.49}$$

$$I \bigcirc F + F \bigcirc I = I_4 \tag{3.3.50}$$

$$\sigma^T(W)\sigma(W) = 2 \tag{3.3.51}$$

$$\sigma^T(W)\sigma(Q) = -(q_{11}+q_{22}) = -\operatorname{tr}(Q) \tag{3.3.52}$$

$$\sigma^T(W)\sigma(\Pi) = -(\Pi_{11}+\Pi_{22}) = -\operatorname{tr}(\Pi). \tag{3.3.53}$$

Hence

$$\Delta(W,F,R,Q) = 1 + 2\operatorname{tr}(Q) . \tag{3.3.54}$$

Theorem 3.3.2 then implies that the solution of (3.3.48) has a finite escape time

(i) for all Π, whenever $tr(Q) < -\frac{1}{2}$

(ii) for $tr(\Pi) < 1 - \sqrt{1+2tr(Q)}$, whenever $tr(Q) \geqslant -\frac{1}{2}$.

From these and Kalman's conditions we can deduce that

(a) when $Q = 0$ the solution exists on $[0,\infty)$ whenever $\Pi \geqslant 0$ and escapes in a finite time whenever $tr(\Pi) < 0$;

(b) when $\Pi = 0$, the solution exists on $[0,\infty)$ whenever $Q \geqslant 0$ and escapes in a finite time whenever $\sqrt{2tr(Q)+1} - 1 < 0$ or $tr(Q) < -\frac{1}{2}$, i.e. $tr(Q) < 0$.

As it is much less restrictive to stipulate that the trace of a matrix should have a definite sign than it is to require the matrix itself to be definite, the finite escape time conditions in (a) and (b) are less restrictive in form than the Kalman existence conditions.

Note that in (ii) above the condition on $tr(\Pi)$ is dependent on $tr(Q)$ so that we cannot state conditions on Π and Q which are independent of one another. This restriction in stating the conditions could be minimized by selecting a W other than I which would in some sense maximize the variety of Q's and Π's which would satisfy the inequality (3.3.45).

3.3.4 Models in Population Dynamics

A popular model for describing the dynamic behaviour of n interacting populations x_i, $i=1,\ldots,n$ is

$$\dot{x}_i = x_i\left(\sum_{j=1}^{n} a_{ij}x_j+\varepsilon_i\right), \quad x_i(0) = x_{io} > 0. \qquad (3.3.55)$$

This set of differential equations can be put in the form of

(3.3.2) by defining the following parameters for $i=1,\ldots,n$

$$A_i = \begin{pmatrix} & & & \tfrac{1}{2}a_{i1} & & & \\ & & & \vdots & & & \\ \tfrac{1}{2}a_{i1} & \cdots & & a_{ii} & \tfrac{1}{2}a_{ii+1} & \cdots & \tfrac{1}{2}a_{in} \\ & & & \tfrac{1}{2}a_{ii+1} & & & \\ & & & \vdots & & & \\ & & & \tfrac{1}{2}a_{in} & & & \end{pmatrix} \qquad (3.3.56)$$

$$b_i^T = (0,\ldots,0,\varepsilon_i,0,\ldots,0) \qquad (3.3.57)$$

$$c_i = 0. \qquad (3.3.58)$$

From (3.3.58) it is clear that $c_\omega = 0$ so that $\Delta_\omega > 0$ owing to the assumed positive definiteness of A_ω. The condition for a finite escape time becomes

$$\omega^T x_0 > -\tfrac{1}{2}\omega^T A_\omega^{-1} b_\omega + \tfrac{1}{2}\sqrt{\omega^T \omega b_\omega^T A_\omega^{-1} b_\omega / \lambda_\omega}\ . \qquad (3.3.59)$$

Furthermore, as $\frac{1}{\lambda_\omega} > 0$ is the largest eigenvalue of A^{-1} we have, by the Cauchy-Schwarz inequality, that

$$\omega^T \omega b_\omega^T (A_\omega^{-1}/\lambda_\omega) b_\omega \geqslant \omega^T \omega b_\omega^T A_\omega^{-1} A_\omega^{-1} b_\omega \geqslant (\omega^T A_\omega^{-1} b_\omega)^2. \qquad (3.3.60)$$

Inequality (3.3.59) thus becomes

$$\omega^T x_0 > 0 \qquad (3.3.61)$$

which, incidentally, is the sufficient condition for a finite escape time valid for all equations (3.3.2) for which $c_\omega = 0$.

3.4 Invariant Sets

In Section 3.3 we obtained sufficient conditions for the existence of finite escape times of solutions of quadratic differential equations for which there exists $\omega \in R^n$ such that

$$\sum_{i=1}^{n} \omega_i A_i > 0. \tag{3.4.1}$$

This is a rather stringent condition which precludes, for example, an analysis of the system

$$\dot{x}_1 = 2x_1x_2 \quad x_1(0) = \varepsilon > 0$$

$$\dot{x}_2 = 2x_1x_2 \quad x_2(0) = 1{,}5\varepsilon > 0 \tag{3.4.2}$$

whose solution obviously has a finite escape time. Therefore instead of insisting upon the existence of $\omega \in R^n$ such that (3.4.1) holds we require only that $x^T(\sum_{i=1}^{n} \omega_i A_i)x$ is definite on a suitable invariant set of (3.3.2). For simplicity we shall assume that $b_i = 0$, $c_i = 0$, $i=1,\ldots,n$ so that (3.3.2) becomes

$$\dot{x}_i = x^T A_i x, \quad x_i(0) = x_{io}, \quad i=1,\ldots,n\ . \tag{3.4.3}$$

LEMMA 3.4.1 The set $\{x \in R^n | Dx \geqslant 0\}$ where $D \in R^{n \times n}$ is an invariant set associated with (3.4.3), that is $x(t)$ lies in this set for all t for which $x(t)$ exists, if

$$Dx_0 > 0 \tag{3.4.4}$$

and

$$x^T\left(\sum_{j=1}^{n} D_{ij}A_j\right)x \geqslant 0 \text{ for all } x \in R^n \text{ such that } Dx \geqslant 0,\ i=1,\ldots,n. \tag{3.4.5}$$

PROOF For each i we have that

$$\frac{d}{dt}\left(\sum_{j=1}^{n} D_{ij}x_j\right) = x^T\left(\sum_{j=1}^{n} D_{ij}A_j\right)x \ . \tag{3.4.6}$$

Now if this time derivative is non-negative for all $x \in R^n$ such that $Dx \geqslant 0$ we have that the components of Dx are monotone non-decreasing in this set. This, together with the fact that $Dx_0 > 0$, implies that $Dx > 0$ for all t for which x exists.

Inequality (3.4.5) concerns the non-negativity of a not necessarily convex quadratic function on a subset of R^n defined by linear inequalities. We therefore use Theorem 3.2.2 to obtain the next lemma.

LEMMA 3.4.2 The set $\{x \in R^n | Dx \geqslant 0\}$ where $D \in R^{n \times n}$ is an invariant set associated with (3.4.3) if

$$Dx_0 > 0 \tag{3.4.7}$$

and there exist copositive matrices $C_i \in R^{n \times n}$ such that for i=1,...,n

$$\sum_{j=1}^{n} D_{ij}A_j - D^TC_iD \geqslant 0 \text{ (positive semi-definite).} \tag{3.4.8}$$

PROOF The proof follows from Lemma 3.4.1 and Theorem 3.2.2.

We can now state the main theorem.

THEOREM 3.4.1 Suppose there exist $k_1 > 0$, $D \in R^{n \times n}$, $C_i \in R^{n \times n}$ and copositive $i=1,\dots,n$, $W \in R^{n \times n}$ and copositive, and $\omega \in R^n$ such that

$$\omega^T x_o > 0 \tag{3.4.9}$$

$$Dx_o > 0 \tag{3.4.10}$$

$$\sum_{j=1}^{n} D_{ij} A_j - D^T C_i D \geqslant 0, \quad i=1,\dots,n \tag{3.4.11}$$

$$\sum_{j=1}^{n} \omega_j A_j - D^T W D - k_1 \omega\omega^T \geqslant 0. \tag{3.4.12}$$

Then, the solution of (3.4.3) has a finite escape time t_e where

$$t_e \leqslant \frac{1}{k_1(\omega^T x_0)} \quad . \tag{3.4.13}$$

PROOF Lemma 3.4.2 states that $\{x \in R^n | Dx \geqslant 0\}$ is an invariant set associated with (3.4.3). We have also that

$$\frac{d}{dt}(\omega^T x) = x^T\left(\sum_{j=1}^{n} \omega_j A_j\right)x \tag{3.4.14}$$

so that

$$\frac{d}{dt}(\omega^T x) \geqslant k_1(\omega^T x)^2 \tag{3.4.15}$$

for all $x \in R^n$ such that $Dx \geqslant 0$ if (3.4.12) holds for some copositive W. Now $\frac{dy}{dt} = k_1y^2$, $y(0) = \omega^T x_o > 0$ has a finite escape time at $\frac{1}{k_1 y(0)}$ so it follows from the preceding inequality that (3.4.3) has a finite escape time t_e which satisfies (3.4.13).

3.4.1 An Example

Consider again equations (3.4.2). Here we have that

$$A_1 = A_2 = \begin{pmatrix} 0 & 1 \\ 1 & 0 \end{pmatrix} \tag{3.4.16}$$

so that there is no linear combination of A_1 and A_2 which is positive definite. Theorem 3.3.1 therefore cannot be used and so we turn to Theorem 3.4.1.

We set

$$k_1 = 1, \quad \omega^T = (1 \quad 0) \tag{3.4.17}$$

$$D = \begin{pmatrix} -1 & 1 \\ 2 & -1 \end{pmatrix}, \quad C_1 = 0, \quad C_2 = W = \begin{pmatrix} 2 & 1 \\ 1 & 0 \end{pmatrix} \tag{3.4.18}$$

so that

$$\omega^T x_o = x_{10} > 0 \tag{3.4.19}$$

$$\sum_{j=1}^{2} D_{1j}A_j - D^T C_1 D = 0 \tag{3.4.20}$$

$$\sum_{j=1}^{2} D_{2j}A_j - D^T C_2 D = \begin{pmatrix} 0 & 1 \\ 1 & 0 \end{pmatrix} - \begin{pmatrix} -1 & 2 \\ 1 & -1 \end{pmatrix} \begin{pmatrix} 2 & 1 \\ 1 & 0 \end{pmatrix} \begin{pmatrix} -1 & 1 \\ 2 & -1 \end{pmatrix}$$

$$= \begin{pmatrix} 2 & 0 \\ 0 & 0 \end{pmatrix} \geqslant 0 \tag{3.4.21}$$

$$\sum_{j=1}^{2} \omega_j A_j - k_1 \omega\omega^T - D^T W D = \begin{pmatrix} 0 & 1 \\ 1 & 0 \end{pmatrix} - \begin{pmatrix} 1 & 0 \\ 0 & 0 \end{pmatrix} - \begin{pmatrix} -2 & 1 \\ 1 & 0 \end{pmatrix}$$

$$= \begin{pmatrix} 1 & 0 \\ 0 & 0 \end{pmatrix} \geqslant 0 \qquad (3.4.22)$$

and

$$Dx_0 = \begin{pmatrix} -1 & 1 \\ 2 & -1 \end{pmatrix} \begin{pmatrix} \varepsilon \\ 1.5\varepsilon \end{pmatrix} = \begin{pmatrix} .5\varepsilon \\ .5\varepsilon \end{pmatrix} > 0 . \qquad (3.4.23)$$

The conditions of Theorem 3.4.1 are therefore satisfied and (3.4.2) has a finite escape time $t_e \leqslant \frac{1}{\varepsilon}$.

3.4.2 General Case

Both Lemma 3.4.2 and Theorem 3.4.1 can be extended to the general quadratic equation (3.3.2), but the results are more cumbersome, especially in so far as a proper generalization of Theorem 3.4.1 is concerned. However the proofs follow easily from an application of the matrix theory of Sections 3.1 and 3.2.

LEMMA 3.4.3 The set $\{x \in R^n | D(x-\tilde{x}) \geqslant 0\}$ is an invariant set associated with (3.3.2) if there exist $C_i \in R^{n \times n}$ and copositive, $\hat{x}_i$, $h_i \in R^n$, i=1,...,n such that

$$2\sum_{j=1}^{n} D_{ij} A_j \hat{x}_i + \sum_{j=1}^{n} D_{ij} b_j + D^T h_i = 0, \quad i=1,\dots,n \qquad (3.4.24)$$

$$h_i \leqslant 0, \quad h_i^T D(\hat{x}_i - \tilde{x}) = 0, \quad D(\hat{x}_i - \tilde{x}) \geqslant 0, \qquad i=1,\dots,n \qquad (3.4.25)$$

$$D(x_0-\tilde{x}) > 0 \tag{3.4.26}$$

$$\sum_{j=1}^{n} D_{ij}A_j - D^T C_i D \geqslant 0, \quad i=1,\ldots,n \tag{3.4.27}$$

$$C_i D(\hat{x}_i-\tilde{x}) = 0, \quad i=1,\ldots,n \tag{3.4.28}$$

and

$$\hat{x}_i^T\left(\sum_{j=1}^{n} D_{ij}A_j\right)\hat{x}_i + \hat{x}_i^T\left(\sum_{j=1}^{n} D_{ij}b_j\right) + \sum_{j=1}^{n} D_{ij}c_j \geqslant 0, \quad i=1,\ldots,n. \tag{3.4.29}$$

PROOF Let D_i denote the i-th row of D. Then

$$\frac{d}{dt}[D_i(x-\tilde{x})] = x^T\left(\sum_{j=1}^{n} D_{ij}A_j\right)x + x^T\left(\sum_{j=1}^{n} D_{ij}b_j\right) + \sum_{j=1}^{n} D_{ij}c_j. \tag{3.4.30}$$

Well-known necessary conditions for the right-hand side to be minimized with respect to x subject to $D(x-\tilde{x}) \geqslant 0$ are that the minimizer $\hat{x}_i$ satisfy (3.4.24), (3.4.25). We prove now that inequalities (3.4.27) and (3.4.28) are then sufficient conditions for $\hat{x}_i$ to be the global minimizer of the right-hand side of (3.4.30).

We define $\phi_i(x)$ as the right-hand side of (3.4.29). Then

$$\phi_i(\hat{x}_i+\delta x) = \phi_i(\hat{x}_i) + \left[2\sum_{i=1}^{n} D_{ij}A_j\hat{x}_i + \sum_{j=1}^{n} D_{ij}b_j\right]^T\delta x + \delta x^T\left(\sum_{j=1}^{n} D_{ij}A_j\right)\delta x \tag{3.4.31}$$

where δx is an arbitrary perturbation which must satisfy

$$D(\hat{x}_i+\delta x-\tilde{x}) \geqslant 0. \tag{3.4.32}$$

Now the second term on the right-hand side of (3.4.31) is non-negative by virtue of (3.4.24) and (3.4.25). Therefore a sufficient condition for $\hat{x}_i$ to be a global minimizer is that the third term on the right-hand side of (3.4.31) should be non-negative for all δx which satisfy (3.4.32). That conditions (3.4.27) and (3.4.28) guarantee this, is proved in Theorem 3.2.3.

Inequality (3.4.29) allows us to conclude that

$$\frac{d}{dt}[D_i(x-\tilde{x})] \geqslant 0, \quad i=1,\ldots,n \tag{3.4.33}$$

and this together with (3.4.26) yields the lemma.

THEOREM 3.4.2 Suppose that $\{x \in R^n | D(x-\tilde{x}) \geqslant 0\}$ is an invariant set associated with (3.3.2), and that there exist $k_1 > 0$, k_2, k_3, $W \in R^{n \times n}$ and copositive, v, $\bar{x}$, $\omega \in R^n$ such that

$$2[\sum_{j=1}^{n} \omega_j A_j - k_1 \omega\omega^T]\bar{x} + \sum_{j=1}^{n} \omega_j b_j - k_2\omega + D^T v = 0 \tag{3.4.34}$$

$$v \leqslant 0, \quad v^T D(\bar{x}-\tilde{x}) = 0, \quad D(\bar{x}-\tilde{x}) \geqslant 0 \tag{3.4.35}$$

$$\sum_{j=1}^{n} \omega_j A_j - k_1 \omega\omega^T - D^T W D \geqslant 0 \tag{3.4.36}$$

$$WD(\bar{x}-\tilde{x}) = 0 \tag{3.4.37}$$

$$\bar{x}^T(\sum_{j=1}^{n} \omega_j A_j - k_1\omega\omega^T)\bar{x} + \bar{x}^T(\sum_{j=1}^{n} \omega_j b_j) + \sum_{j=1}^{n} \omega_j c_j - k_2\bar{x}^T\omega - k_3 \geqslant 0. \tag{3.4.38}$$

Then, the solution of (3.3.2) has a finite escape time according to conditions (i) and (ii) of Lemma 3.3.1, and Lemma 3.3.2, where y_0 is set as $\omega^T x_0$.

PROOF It is easily verified that conditions (3.4.34) - (3.4.38) imply that

$$\frac{d}{dt}(\omega^T x) \geqslant k_1(\omega^T x)^2 + k_2(\omega^T x) + k_3 \qquad (3.4.39)$$

for all $x \in \{x \in R^n | D(x-\tilde{x}) \geqslant 0\}$, which allows the use of Lemmas 3.3.1 and 3.3.2.

3.5 Conclusion

In this chapter we presented a number of matrix-theoretic results relating to non-convex quadratic forms. In particular, copositive matrices and their role in quadratic programming were emphasized. An extension of Finsler's theorem was also presented.

Quadratic differential equations were studied with a view to obtaining conditions for the existence of finite escape times in their solutions. On the assumption that a linear combination of certain matrices is positive definite, a set of conditions was obtained which was applied to the matrix Riccati and certain population equations.

The notion of invariant sets was introduced and exploited to relax the restrictive positive definiteness requirement on the linear combination of the system matrices. Here too copositive matrices played an important role.

In Chapter 4 we turn to the development of conditions for the non-negativity of non-quadratic functionals, thus generalizing certain of the results of Chapter 4 of [23].

3.6 References

[1] DIANANDA, P.H. On Non-negative Forms in Real Variables Some or All of which are Non-negative. Proc. Cambridge Philos. Soc., 58, 1962, pp. 17-25.

[2] GADDUM, J.W. Linear Inequalities and Quadratic Forms. Pacific J. Math., 8, 1958, pp. 411-414.

[3] HALL, M. & NEWMAN, M. Copositive and Completely Positive Quadratic Forms. Proc. Cambridge Philos. Soc., 59, 1963, pp. 329-339.

[4] BAUMERT, L.D. Extreme Copositive Quadratic Forms. Pacific J. Math., 19, 1966, pp. 197-204.

[5] BAUMERT, L.D. Extreme Copositive Quadratic Forms II. Pacific J. Math., 20, 1967, pp. 1-20.

[6] MARKHAM, T.L. Factorizations of Completely Positive Matrices. Proc. Cambridge Philos. Soc., 69, 1971, pp. 53-58.

[7] COTTLE, R.W. & MYLANDER, W.C. Ritter's Cutting Plane Method for Non-convex Quadratic Programming. In: Integer and Non-linear Programming, edited by J. Abadie, 1970, pp. 257-283.

[8] COTTLE, R.W., HABETLER, G.J. & LEMKE, C.E. Quadratic Forms Semi-definite over Convex Cones. Proceedings of the Princeton Symposium on Mathematical Programming, edited by H.W. Kuhn, Princeton University Press, 1970, pp. 551-565.

[9] COTTLE, R.W. On the Convexity of Quadratic Forms over Convex Sets. Operations Research, 15, 1967, pp. 170-172.

[10] COTTLE, R.W. Note on a Fundamental Theorem in Quadratic Programming. SIAM J. Applied Mathematics, 12, 1964, pp. 663-665.

[11] MARTOS, B. Quadratic Programming with a Quasiconvex Objective Function. Operations Research, 19, 1971, pp. 87-97.

[12] COTTLE, R.W. & FERLAND, J.A. On Pseudo-convex Functions of Non-negative Variables. Math. Programming, 1, 1971, pp. 95-101.

[13] KARAMARDIAN, S. The Complementary Problem. Math. Programming, 2, 1972, pp. 107-129.

[14] JACOBSON, D.H. Factorization of Symmetric M-Matrices. Linear Algebra and its Applications, 9, 1974, pp. 275-278.

[15] FINSLER, P. Über das Vorkommen definiter und semi-definiter Formen in Scharen quadratischer Formen. Commentarii Mathematici Helvetici, 9, 1937, pp. 188-192.

[16] JACOBSON, D.H. A Generalization of Finsler's Theorem for Quadratic Inequalities and Equalities. Quaestiones Mathematicae, 1, 1976, pp. 19-28.

[17] BELLMAN, R. Introduction to Matrix Analysis. McGraw-Hill, New York, 1970.

[18] LA SALLE, J.P. & LEFSCHETZ, S. Stability by Liapunov's Direct Method with Applications. Academic Press, New York, 1961, Chapter 4.

[19] FRAYMAN, M. Quadratic Differential Systems: A Study in Non-linear Systems Theory. Ph.D. thesis, Department of Electrical Engineering, University of Maryland, 1974.

[20] JACOBSON, D.H. Conditions for Existence of Finite Escape Times and Divergence of Solutions of Quadratic Differential Equations. Internal Report, Department of Applied Mathematics, University of the Witwatersrand, Johannesburg, Republic of South Africa, 1972, 20 pages.

[21] GETZ, W.M. & JACOBSON, D.H. Sufficiency Conditions for Finite Escape Times in Systems of Quadratic Differential Equations. J. Inst. Math. Applic., to appear.

[22] JACOBSON, D.H. New Conditions for Boundedness of the Solution of a Matrix Riccati Differential Equation. J. Differential Equations, 8, 1970, pp. 258-263.

[23] BELL, D.J. & JACOBSON, D.H. Singular Optimal Control Problems. Academic Press, New York and London, 1975.

4. NON-NEGATIVITY CONDITIONS FOR CONSTRAINED AND NON-QUADRATIC FUNCTIONALS

4.1 Linear-quadratic Case

Necessary and sufficient conditions for the non-negativity of quadratic functionals have been derived during the past several years [1] - [4] and have found application in singular optimal control and estimation theory [5]. Chapter 4 of [1] and [2] - [4] together provide a complete treatment of the problem of deciding whether the quadratic performance criterion

$$J[u(\cdot)] = \int_0^T (\tfrac{1}{2}x^T Qx+u^T Cx+\tfrac{1}{2}u^T Ru)dt + \tfrac{1}{2}x^T(T)Q_T x(T) \quad (4.1.1)$$

is non-negative for all piecewise continuous functions $u(\cdot)$ where

$$\dot{x} = Ax + Bu \ , \quad x(0) = x_0 = 0 \quad (4.1.2)$$

$$Dx(T) = 0 \quad (4.1.3)$$

and where the matrices $Q \in R^{n \times n}$, $C \in R^{m \times n}$, $R \in R^{m \times m}$, $A \in R^{n \times n}$ and $B \in R^{n \times m}$ are assumed to be continuous functions of time on $[0,T]$. The matrices $Q_T \in R^{n \times n}$ and $D \in R^{s \times n}$ are constant. With no loss of generality the matrices Q, R and Q_T are assumed to be symmetric and D is assumed to have rank s.

In the following pages we present certain results on the above-mentioned linear-quadratic problem and interpret these in new ways which enable us to obtain sufficient conditions for non-negativity of functionals which are non-linear-quadratic.

4.1.1 Non-negativity Conditions

A classical necessary condition for non-negativity of $J[u(\cdot)]$ is provided by the first theorem.

THEOREM 4.1.1 A necessary condition for $J[u(\cdot)]$ to be non-negative for all piecewise continuous control functions $u(\cdot)$ which cause (4.1.3) to be satisfied is that

$$R(t) \geqslant 0 \text{ for all } t \in [0,T]. \tag{4.1.4}$$

In the Calculus of Variations condition (4.1.4) is referred to as the Legendre-Clebsch necessary condition.

In fact, an even stronger statement is possible, as is confirmed by the next theorem.

THEOREM 4.1.2 A necessary condition for $J[u(\cdot)] \geqslant k\|u(\cdot)\|^2$, $k > 0$ (i.e. for $J[u(\cdot)]$ to be strongly positive) where $\|u(\cdot)\|^2 \triangleq \int_0^T u^T(t)u(t)dt$ is that

$$R(t) > 0 \text{ for all } t \in [0,T]. \tag{4.1.5}$$

Note that if (4.1.5) holds, the functional (4.1.1) is referred to as non-singular, while if (4.1.4) holds, it is referred to as partially singular. In the non-singular case we have the following well-known result which is proved in [1].

THEOREM 4.1.3 A necessary and sufficient condition for $J[u(\cdot)]$ to be strongly positive when $D \equiv 0$ is that (4.1.5) holds and that there exists for all $t \in [0,T]$ a symmetric matrix function of time $S(\cdot)$ which satisfies the matrix Riccati differential equation

$$- \dot{S} = Q + SA + A^T S - (C+B^T S)^T R^{-1}(C+B^T S) \qquad (4.1.6)$$

$$S(T) = Q_T. \qquad (4.1.7)$$

In the singular case (4.1.6) fails, owing to the non-invertibility of R. However, the following sufficient condition is valid.

THEOREM 4.1.4 [6] A sufficient condition for $J[u(\cdot)]$ to be non-negative for all piecewise continuous functions $u(\cdot)$ is that there exists for all $t \in [0,T]$ a continuously differentiable, symmetric matrix function of time $P(\cdot)$ such that

$$\begin{pmatrix} \dot{P} + Q + PA + A^T P & (C+B^T P)^T \\ C + B^T P & R \end{pmatrix} \geqslant 0 \qquad (4.1.8)$$

for all $t \in [0,T]$ and

$$Z^T [Q_T - P(T)] Z \geqslant 0 \qquad (4.1.9)$$

where

$$Z = \begin{pmatrix} -D_1^{-1} D_2 \\ I_{n-s} \end{pmatrix}, \; D = [D_1 \;\; D_2] \;, \; D_1 \in R^{s \times s}, \; D_2 \in R^{s \times (n-s)} . \qquad (4.1.10)$$

Note that the assumed rank condition on D guarantees that it is always possible to re-label variables so that D_1 is invertible. In the totally singular case ($R \equiv 0$) we have that condition (4.1.8) becomes

$$\dot{P} + Q + PA + A^T P \geqslant 0 \qquad (4.1.11)$$

$$C + B^T P = 0 \qquad (4.1.12)$$

for all $t \in [0,T]$.

Though the above conditions (4.1.8)-(4.1.12) are not in general necessary, they differ from necessary and sufficient conditions only in small detail [1], [7]. As the results to be presented in this chapter are mainly of a sufficiency nature, we prefer to state Theorem 4.1.4 rather than the more abstract necessary and sufficient conditions.

The following theorem relates (4.1.11), (4.1.12) to a matrix Riccati equation.

THEOREM 4.1.5 [1] Suppose that $R \equiv 0$ for all $t \in [0,T]$ and that (generalized Legendre-Clebsch condition)

$$(-1) \frac{\partial}{\partial u} \ddot{H}_u > 0 \qquad (4.1.13)$$

$$\frac{\partial}{\partial u} \dot{H}_u = 0 \qquad (4.1.14)$$

for all $t \in [0,T]$ where

$$H \triangleq \tfrac{1}{2} x^T Q x + u^T C x + \lambda^T (Ax+Bu) \qquad (4.1.15)$$

and

$$-\dot{\lambda} \triangleq Qx + A^T \lambda + C^T u. \qquad (4.1.16)$$

Suppose further that there exists a function $S(\cdot)$ which satisfies for all $t \in [0,T]$ the matrix Riccati equation

$$- \dot{S} = Q + SA + A^T S$$

$$+ [(AB-\dot{B})^T S + B^T Q - CA - \dot{C}]^T [\frac{\partial}{\partial u} \ddot{H}_u]^{-1} [(AB-\dot{B})^T S + B^T Q - CA - \dot{C}] \tag{4.1.17}$$

$$C(T) + B^T(T)S(T) = 0 \tag{4.1.18}$$

$$Z^T [Q_T - S(T)] Z \geqslant 0. \tag{4.1.19}$$

Then there exists a $P(\cdot)$ which satisfies (4.1.11), (4.1.12) and (4.1.9). Note that it is easy to prove that the conditions of Theorem 4.1.3 imply satisfaction of those of Theorem 4.1.4.

The results summarized above indicate the close relationship between the existence of a matrix function of time which satisfies certain matrix differential and algebraic inequalities and equalities, and the non-negativity of $J[u(\cdot)]$. It is this relationship which we exploit later in this chapter to develop sufficiency conditions for the non-negativity of non-linear-quadratic functionals.

Our next observation concerning the preceding theorems is also of some importance. Let us define

$$\bar{M} = \begin{pmatrix} \bar{M}_{11} & \bar{M}_{21}^T \\ \bar{M}_{21} & \bar{M}_{22} \end{pmatrix} \triangleq \begin{pmatrix} \dot{P} + Q + PA + A^T P & (C+B^T P)^T \\ C + B^T P & R \end{pmatrix} \tag{4.1.20}$$

$$\bar{N} \triangleq Z^T [Q_T - P(T)] Z. \tag{4.1.21}$$

We can then state and prove the following theorem.

THEOREM 4.1.6 Suppose that the conditions of Theorem 4.1.4 are satisfied. Then there exist a continuous, positive semi-definite symmetric matrix function of time $M(\cdot)$ and a positive semi-definite symmetric matrix N such that

$$J[u(\cdot)] \geqslant \int_0^T (\tfrac{1}{2}x^T M_{11}x+u^T M_{21}x+\tfrac{1}{2}u^T M_{22}u)dt + \tfrac{1}{2}y^T(T)Ny(T) \tag{4.1.22}$$

for all $u(\cdot)$ such that (4.1.3) holds, where

$$y^T(T) \triangleq (x_{s+1}(T),\ldots,x_n(T)). \tag{4.1.23}$$

PROOF Add the identically zero integral $\int_0^T \tfrac{1}{2}x^T P(Ax+Bu-\dot{x})dt$ to $J[u(\cdot)]$ and integrate by parts to obtain

$$J[u(\cdot)] = \int_0^T [\tfrac{1}{2}x^T(\dot{P}+Q+PA+A^TP)x+u^T(C+B^TP)x+\tfrac{1}{2}u^TRu]dt$$

$$+ \tfrac{1}{2}x_0^TP(0)x_0 + \tfrac{1}{2}x^T(T)[Q_T-P(T)]x(T). \tag{4.1.24}$$

Now because $u(\cdot)$ is chosen so that (4.1.3) holds, we have that

$$x(T) = Zy(T) \tag{4.1.25}$$

where $y(T)$ is defined by (4.1.23).

As a consequence of the fact that $x_0 = 0$, and using (4.1.20), (4.1.21) $J[u(\cdot)]$ becomes

$$J[u(\cdot)] = \int_0^T \tfrac{1}{2}[x^T \; u^T]\bar{M}[x^T \; u^T]^T dt + \tfrac{1}{2}y^T(T)\bar{N}y(T) \tag{4.1.26}$$

which is (4.1.22) holding with equality.

Note that the fact that $J[u(\cdot)]$ is strongly positive if and only if the conditions of Theorem 4.1.3 hold, implies that (4.1.22) holds with $M_{11} = 0$, $M_{21} = 0$, $M_{22} = kI$ for some $k > 0$, $N = 0$. Consequently the sufficient condition presented in the next theorem is implied by the known sufficient conditions for the non-singular, partially singular and singular linear-quadratic cases.

THEOREM 4.1.7 A sufficient condition for $J[u(\cdot)] \geqslant 0$ for all $u(\cdot)$ for which (4.1.3) holds is that there exist a continuous, positive semi-definite symmetric matrix function of time $M(\cdot)$ and a positive semi-definite symmetric matrix N such that

$$J[u(\cdot)] \geqslant \int_0^T (\tfrac{1}{2}x^T M_{11} x + u^T M_{21} x + \tfrac{1}{2}u^T M_{22} u)dt + \tfrac{1}{2}y^T(T)Ny(T) \tag{4.1.27}$$

for all $u(\cdot)$ for which (4.1.3) holds.

4.1.2 Novel Non-negativity Conditions

First we note that $\bar{M} \geqslant 0$ if and only if the following conditions hold. Either

$$\bar{M}_{11} \geqslant 0 \tag{4.1.28}$$

$$\bar{M}_{22} \geqslant 0 \tag{4.1.29}$$

$$N(\bar{M}_{21}) \supseteq N(\bar{M}_{11}) \tag{4.1.30}$$

$$\bar{M}_{22} - \bar{M}_{21}\bar{M}_{11}^{+}\bar{M}_{21}^{T} \geqslant 0 \tag{4.1.31}$$

where '+' denotes pseudo-inverse, or

$$\bar{M}_{11} \geqslant 0 \tag{4.1.32}$$

$$\bar{M}_{22} \geqslant 0 \tag{4.1.33}$$

$$N(\bar{M}_{21}^T) \supseteq N(\bar{M}_{22}) \tag{4.1.34}$$

$$\bar{M}_{11} - \bar{M}_{21}^T \bar{M}_{22}^+ \bar{M}_{21} \geqslant 0. \tag{4.1.35}$$

If we use the second set of conditions when $R > 0$ we set

$$\bar{M}_{22} = R > 0 \tag{4.1.36}$$

$$\bar{M}_{21} = (C+B^T P) \tag{4.1.37}$$

$$\bar{M}_{11} = (C+B^T P)^T R^{-1} (C+B^T P) \tag{4.1.38}$$

which clearly satisfy (4.1.34), and result in the well-known Riccati equation

$$-\dot{P} = Q + PA + A^T P - (C+B^T P)^T R^{-1} (C+B^T P) \tag{4.1.39}$$

$$z^T [Q_T - P(T)] z \geqslant 0 . \tag{4.1.40}$$

In the partially-singular case we could hope to replace R^{-1} by R^+ but we have no cause to suspect that (4.1.34) will be satisfied. In fact, in general it will not be satisfied; in particular, in the totally singular case the correct form for $\bar{M}_{11}$ is given in (4.1.17).

However we can state and prove the following theorems.

THEOREM 4.1.8 $J[u(\cdot)]$ is non-negative for all $u(\cdot)$ for which (4.1.3) holds if there exist for all $t \in [0,T]$ a continuously differentiable symmetric matrix function of time $P(\cdot)$ and a positive definite symmetric matrix function of time $W(\cdot)$ such that

$$-\dot{P} = Q + PA + A^T P - \bar{M}_{11} \tag{4.1.41}$$

$$Z^T [Q_T - P(T)] Z \geqslant 0 \tag{4.1.42}$$

$$R - (C+B^T P)\bar{M}_{11}^{+}(C+B^T P)^T \geqslant 0 \text{ for all } t \in [0,T] \tag{4.1.43}$$

where

$$\bar{M}_{11} = (C+B^T P)^T W^{-1}(C+B^T P). \tag{4.1.44}$$

PROOF Here $\bar{M}_{11}$ is given by (4.1.44) and $\bar{M}_{21} = C + B^T P$, $\bar{M}_{22} = R$. With this choice, conditions (4.1.28)-(4.1.31) are satisfied.

THEOREM 4.1.9 If $R \geqslant W$ inequality (4.1.43) is always satisfied.

PROOF First we note that $\bar{M}_{11}$ can be written as

$$\bar{M}_{11} = [\tilde{W}(C+B^T P)]^T [\tilde{W}(C+B^T P)] \tag{4.1.45}$$

where the symmetric matrix $\tilde{W}$ is defined by

$$\tilde{W} \triangleq W^{-\frac{1}{2}}. \tag{4.1.46}$$

The pseudo-inverse $\bar{M}_{11}^{+}$ is then

$$[\tilde{W}(C+B^TP)]^+[\tilde{W}(C+B^TP)]^{T^+} \qquad (4.1.47)$$

so that the left-hand side of (4.1.43) is

$$R - (C+B^TP)[\tilde{W}(C+B^TP)]^+[\tilde{W}(C+B^TP)]^{T^+}(C+B^TP)^T. \quad (4.1.48)$$

Pre- and post-multiplying (4.1.48) by $\tilde{W}$ the symmetric square root of W^{-1}, yields

$$\tilde{W}R\tilde{W} - [\tilde{W}(C+B^TP)][\tilde{W}(C+B^TP)]^+[\tilde{W}(C+B^TP)]^{T^+}[\tilde{W}(C+B^TP)]^T \qquad (4.1.49)$$

which is

$$\tilde{W}R\tilde{W} - [\tilde{W}(C+B^TP)][\tilde{W}(C+B^TP)]^+. \qquad (4.1.50)$$

Setting $R = W + \phi$, where ϕ is positive semi-definite yields

$$I + \tilde{W}\phi\tilde{W} - [\tilde{W}(C+B^TP)][\tilde{W}(C+B^TP)]^+ \qquad (4.1.51)$$

which, by the properties of pseudo-inverses and the symmetry of $\tilde{W}$ is positive semi-definite. Pre-multiplying and post-multiplying (4.1.51) by $\tilde{W}^{-1}$ preserves this positive semi-definiteness, and (4.1.43) follows.

Theorem 4.1.8 can be generalized via (4.1.28)-(4.1.31) to read as follows.

THEOREM 4.1.10 $J[u(\cdot)]$ is non-negative for all $u(\cdot)$ for which (4.1.3) holds if there exist for all $t \in [0,T]$ a continuously differentiable symmetric matrix function of time $P(\cdot)$ and a positive semi-definite symmetric matrix function of time $\bar{M}_{11}$ such that

$$- \dot{P} = Q + PA + A^T P - \bar{M}_{11} \tag{4.1.52}$$

$$Z^T [Q_T - P(T)] Z \geqslant 0 \tag{4.1.53}$$

$$N(C+B^T P) \supseteq N(\bar{M}_{11}) \tag{4.1.54}$$

$$R - (C+B^T P)\bar{M}_{11}^{+}(C+B^T P)^T \geqslant 0 \ . \tag{4.1.55}$$

4.1.3 Example

Suppose that A, B, C, Q, R are constant matrices and that there exist constant symmetric matrices P, $\bar{M}_{11}$ such that

$$Q + PA + A^T P = \bar{M}_{11} > 0. \tag{4.1.56}$$

Such a P will exist for arbitrary Q, $\bar{M}_{11}$ if and only if

$$\lambda_i(A) + \lambda_j(A) \neq 0 \ , \quad i,j=1,\ldots,n \tag{4.1.57}$$

where $\lambda_i(A)$ denotes an eigenvalue of A. Condition (4.1.54) is automatically satisfied owing to the assumed positive-definiteness of $\bar{M}_{11}$.

Sufficient conditions for $J[u(\cdot)] \geqslant 0$ for all $u(\cdot)$ such that (4.1.3) holds are therefore that there exists a constant symmetric matrix P such that

$$Z^T [Q_T - P] Z \geqslant 0 \tag{4.1.58}$$

$$Q + PA + A^T P > 0 \tag{4.1.59}$$

$$R - (C+B^TP)(Q+PA+A^TP)^{-1}(C+B^TP)^T \geqslant 0. \qquad (4.1.60)$$

4.2 Constrained Case

4.2.1 Formulation

We suppose here the same formulation as in Section 4.1 but that the (n+m)-vector $(x^T\ u^T)^T$ is constrained to belong to $\Omega_{xu} \subseteq R^{n+m}$. Specifically, we wish to investigate the non-negativity of

$$J[u(\cdot)] = \int_0^T (\tfrac{1}{2}x^TQx+u^TCx+\tfrac{1}{2}u^TRu)dt + \tfrac{1}{2}x^T(T)Q_Tx(T) \qquad (4.2.1)$$

subject to

$$\dot{x} = Ax + Bu \;, \quad x(0) = 0 \qquad (4.2.2)$$

$$Dx(T) = 0 \qquad (4.2.3)$$

and

$$\begin{pmatrix} x(t) \\ u(t) \end{pmatrix} \varepsilon\ \Omega_{xu} \subseteq R^{n+m} \qquad (4.2.4)$$

where the control function $u(\cdot)$ is admissible if it is piece-wise continuous on [0,T]. Naturally we assume that there exists an admissible control function such that (4.2.3) and (4.2.4) are satisfied.

4.2.2 Sufficiency Conditions

Our first sufficient condition is an immediate generalization of Theorem 4.1.4.

THEOREM 4.2.1 A sufficient condition for $J[u(\cdot)]$ to be non-negative for all piecewise continuous control functions $u(\cdot)$ for which (4.2.3) and (4.2.4) hold is that there exists for all $t \in [0,T]$ a continuously differentiable, symmetric matrix function of time $P(\cdot)$ such that

$$[\xi^T \quad \eta^T]\begin{pmatrix}\dot{P} + Q + PA + A^TP & (C+B^TP)^T \\ C + B^TP & R\end{pmatrix}\begin{pmatrix}\xi \\ \eta\end{pmatrix} \geqslant 0 \tag{4.2.5}$$

for all $\begin{pmatrix}\xi \\ \eta\end{pmatrix} \in \Omega_{xu}$, for all $t \in [0,T]$, and

$$z^T[Q_T-P(T)]z \geqslant 0. \tag{4.2.6}$$

PROOF The proof is very similar to that of Theorem 4.1.6 and we refrain from presenting it here.

Although Theorem 4.2.1 is not easy to use, in that a constructive method for generating $P(\cdot)$ is not suggested, it is important because the constraint set Ω_{xu} is arbitrary. Before going on to consider some special constraint sets which do permit the development of constructive theorems, we state an appropriate generalization of Theorem 4.1.7.

THEOREM 4.2.2 A sufficient condition for $J[u(\cdot)] \geqslant 0$ for all $u(.)$ for which (4.2.3), (4.2.4) hold, is that

$$J[u(\cdot)] \geqslant \int_0^T \tfrac{1}{2}[x^T \; u^T]M[x^T \; u^T]^T dt + \tfrac{1}{2}y^T(T)Ny(T) \tag{4.2.7}$$

where

$$N \geqslant 0 \tag{4.2.8}$$

and the symmetric matrix function of time M(·) satisfies

$$[\xi^T \ \eta^T] \ M\begin{pmatrix}\xi \\ \eta\end{pmatrix} \geqslant 0 \text{ for all } \begin{pmatrix}\xi \\ \eta\end{pmatrix} \varepsilon \ \Omega_{xu} \text{ , for all } t \ \varepsilon \ [0,T] \ . \tag{4.2.9}$$

The first special constraint set which we consider is

$$\Omega_{xu} \equiv \Omega_x x R^m \ , \quad \text{where } \Omega_x \subseteq R^n \tag{4.2.10}$$

or

$$x(t) \ \varepsilon \ \Omega_x \ , \quad u(t) \ \varepsilon \ R^m \ , \quad \text{for all } t \ \varepsilon \ [0,T] \ . \tag{4.2.11}$$

THEOREM 4.2.3 A sufficient condition for $J[u(\cdot)] \geqslant 0$ for all $u(\cdot)$ for which (4.2.3), (4.2.11) hold, is that there exists for all $t \ \varepsilon \ [0,T]$ a continuously differentiable symmetric matrix function of time $P(\cdot)$ such that

$$R \geqslant 0 \tag{4.2.12}$$

$$(C+B^T P)\xi \ \varepsilon \ \mathcal{R}(R) \text{ for all } \xi \ \varepsilon \ \Omega_x \tag{4.2.13}$$

$$\xi^T [\dot{P}+Q+PA+A^T P-(C+B^T P)^T R^+ (C+B^T P)]\xi \geqslant 0 \text{ for all } \xi \ \varepsilon \ \Omega_x \tag{4.2.14}$$

$$Z^T [Q_T - P(T)] Z \geqslant 0. \tag{4.2.15}$$

PROOF As η is unconstrained we can minimize the left-hand side of (4.2.5) with respect to η, provided that (4.2.12) and (4.2.13) are satisfied. This yields (4.2.14) and the theorem is proved.

The corresponding case when

$$x(t) \in R^n,\ u(t) \in \Omega_u \subseteq R^m \text{ for all } t \in [0,T] \qquad (4.2.16)$$

is covered by the following theorem which may be proved in a similar way to Theorem 4.2.3.

THEOREM 4.2.4 A sufficient condition for $J[u(\cdot)] \geqslant 0$ for all $u(\cdot)$ for which (4.2.3), (4.2.16) hold, is that there exists for all $t \in [0,T]$ a continuously differentiable symmetric matrix function of time $P(\cdot)$ such that

$$\bar{M}_{11} \geqslant 0 \qquad (4.2.17)$$

$$(C+B^TP)^T\eta \in \mathcal{R}(\bar{M}_{11}) \text{ for all } \eta \in \Omega_u \qquad (4.2.18)$$

$$\eta^T[R-(C+B^TP)\bar{M}_{11}^{+}(C+B^TP)^T]\eta \geqslant 0 \text{ for all } \eta \in \Omega_u \qquad (4.2.19)$$

$$z^T[Q_T-P(T)]z \geqslant 0 \qquad (4.2.20)$$

where

$$\bar{M}_{11} \triangleq \dot{P} + Q + PA + A^TP. \qquad (4.2.21)$$

Perhaps the most useful special case is that where

$$\begin{pmatrix} x \\ u \end{pmatrix} \in \Omega_{xu} \text{ for all } t \in [0,T] \qquad (4.2.22)$$

where

$$\Omega_{xu} = \left\{ \begin{pmatrix} x \\ u \end{pmatrix} \in R^{n+m} \;\middle|\; F\begin{pmatrix} x \\ u \end{pmatrix} \geqslant 0,\ F \in R^{q\times(n+m)} \right\}. \qquad (4.2.23)$$

In this case we succeed in constructing $P(\cdot)$.

THEOREM 4.2.5 A sufficient condition for $J[u(\cdot)] \geq 0$ for all $u(\cdot)$ for which (4.2.3), (4.2.23) hold, is that there exist for all $t \in [0,T]$ a continuously differentiable symmetric matrix function of time $P(\cdot)$ and a continuous symmetric matrix function of time $G(\cdot)$ such that

$$\begin{pmatrix} \dot{P} + Q + PA + A^T P & (C+B^T P)^T \\ C + B^T P & R \end{pmatrix} - F^T G F \geq 0 \quad (4.2.24)$$

$$Z^T [Q_T - P(T)] Z \geq 0 \quad (4.2.25)$$

where $G(t)$ is copositive, $t \in [0,T]$.

COROLLARY 4.2.1 Suppose we assume that F is block diagonal and we partition the matrix F and the copositive matrix G as

$$F = \begin{pmatrix} F_{11} & 0 \\ 0 & F_{22} \end{pmatrix}, \quad G = \begin{pmatrix} G_{11} & G_{21}^T \\ G_{21} & G_{22} \end{pmatrix} \quad (4.2.26)$$

where $F_{11}, G_{11} \in R^{n \times n}$ and $F_{22}, G_{22} \in R^{m \times m}$, and suppose that for all $t \in [0,T]$

$$R - F_{22}^T G_{22} F_{22} > 0. \quad (4.2.27)$$

Then the conditions of Theorem 4.2.5 are satisfied if there exists for all $t \in [0,T]$ a solution to the matrix Riccati differential equation

$$- \dot{S} = Q - F_{11}^T G_{11} F_{11} + SA + A^T S$$
$$- (C+B^T P-F_{22}^T G_{21} F_{11})^T (R-F_{22}^T G_{22} F_{22})^{-1} (C+B^T P-F_{22}^T G_{21} F_{11}) \tag{4.2.28}$$

$$Z^T [Q_T-P(T)] Z \geq 0. \tag{4.2.29}$$

PROOF From (4.2.24) we see that the copositivity of G implies that

$$[\xi^T \ \eta^T] \begin{bmatrix} \dot{P} + Q + PA + A^T P & (C+B^T P)^T \\ C + B^T P & R \end{bmatrix} \begin{pmatrix} \xi \\ \eta \end{pmatrix}$$

$$\geq [\xi^T \ \eta^T] F^T G F \begin{pmatrix} \xi \\ \eta \end{pmatrix} \geq 0 \tag{4.2.30}$$

for all $\begin{pmatrix} \xi \\ \eta \end{pmatrix} \varepsilon \, \Omega_{xu}$ defined by (4.2.23), and the conditions of Theorem 4.2.1 are satisfied. The proof of Corollary 4.2.1 is straightforward and follows that of Theorem 4.3 of [1].

Theorem 4.2.5 and Corollary 4.2.1 are important in that they answer the question as to whether or not $J[u(\cdot)]$ is non-negative subject to linear constraints on x and u, at least as far as sufficiency is concerned. Conditions for this case which are both necessary and sufficient are not yet available and it seems that their discovery will be a non-trivial task. In particular, necessity usually requires an assumption on the controllability of the dynamic system [1] and this is a complex concept when constraints are present - see Chapter 5.

Theorem 4.2.6 is a direct consequence of Corollary 4.2.1 and Theorem 4.1.3.

THEOREM 4.2.6 A (necessary and) sufficient condition for $J[u(\cdot)] - \int_0^T [x^T \; u^T] F^T G F \begin{bmatrix} x \\ u \end{bmatrix} dt$ to be strongly positive is that there exists a symmetric matrix function of time $P(\cdot)$ such that (4.2.28), (4.2.29) are satisfied.

4.2.3 Induced State Constraints

A control constraint set Ω_u may <u>induce</u> a state constraint set Ω_x even though there is no explicit constraint on the state. For example, suppose we have the system

$$\dot{x} = Ax + Bu \; , \quad x(0) = 0 \qquad (4.2.31)$$

where $A_{ij} \geqslant 0$, $i \neq j$ and $B_{ij} \geqslant 0$ for all i,j and where $\Omega_u = \{u \in R^m | u \geqslant 0\}$. In this case it is well known that $x(t) \geqslant 0$ for all $t \in [0,T]$ and so the state constraint $\Omega_x = \{x \in R^n | x \geqslant 0\}$ is induced by Ω_u and the dynamic system.

A generalization of this result is provided by the following theorem.

THEOREM 4.2.7 Consider the linear dynamic system

$$\dot{x} = Ax + Bu \; , \quad x(0) = 0 \qquad (4.2.32)$$

where $u(t) \in \Omega_u = \{u \in R^m | F_{22} u \geqslant 0\}$.

Suppose that there exists a constant $n \times n$ matrix F_{11} such that

$$(F_{11} A F_{11}^{-1})_{ij} \geqslant 0 \; , \quad i \neq j \; , \quad \text{for all } t \in [0,T] \qquad (4.2.33)$$

and

$$F_{11}Bu \geqslant 0 \text{ for all } u \in \Omega_u, \text{ for all } t \in [0,T]. \quad (4.2.34)$$

Then the state $x(t) \in \Omega_x$ for all $t \in [0,T]$ where the induced constraint set Ω_x is given by

$$\Omega_x = \{x \in R^n | F_{11}x \geqslant 0\}. \quad (4.2.35)$$

PROOF Defining $y = F_{11}x$, we see that

$$\dot{y} = F_{11}AF_{11}^{-1}y + F_{11}Bu \ , \quad y(0) = 0 \quad (4.2.36)$$

and $y \geqslant 0$ if (4.2.33) and (4.2.34) hold.

4.3 Non-linear Systems

The results of Section 4.2 followed rather easily from those of Section 4.1 for the unconstrained linear-quadratic formulation. The question that arises now is whether or not similar extensions are possible to (constrained) non-linear systems. Some results which hint that this may indeed be feasible are contained in [8], [9]. In particular in [8] the controllability of the non-linear system

$$\dot{x} = A(x,t)x + B(x,t)u \quad (4.3.1)$$

is related to that of the linear time-varying system

$$\dot{x} = A(y,t)x + B(y,t)u \quad (4.3.2)$$

where $y(\cdot)$ belongs to a certain set of continuous vector functions. Loosely speaking, the sufficiency result is obtained that (4.3.1) is 'controllable' if (4.3.2) is 'controllable' for all continuous $y(\cdot)$.

It turns out that we can indeed make statements about the non-negativity of non-quadratic functionals subject to the dynamical constraint (4.3.1) by considering a related (constrained) linear-quadratic problem.

4.3.1 Formulation

We consider the question of whether or not the functional

$$J[u(\cdot)] = \int_0^T [\tfrac{1}{2}x^T Q(x,u,t)x + u^T C(x,u,t)x + \tfrac{1}{2}u^T R(x,u,t)u]\,dt$$

$$+ \tfrac{1}{2}x^T(T)Q_T(x(T))x(T) \tag{4.3.3}$$

is non-negative subject to the constraints

$$\dot{x} = A(x,u,t)x + B(x,u,t)u\ , \quad x(0) = 0 \tag{4.3.4}$$

$$D(x(T))x = 0 \tag{4.3.5}$$

$$\begin{pmatrix} x \\ u \end{pmatrix} \varepsilon\ \Omega_{xu} \subseteq R^{n+m} \text{ for all } t\ \varepsilon\ [0,T] \tag{4.3.6}$$

and where $u(\cdot)$ is piecewise continuous on $[0,T]$. The matrices A, B, Q, C, R are assumed to be continuous functions of their arguments. Furthermore Q, R and Q_T are assumed to be symmetric for all values of their arguments.

The dynamic system (4.3.4) is assumed to have a solution defined on $[0,T]$ for each piecewise continuous input $u(\cdot)$.

4.3.2 Non-negativity Conditions

We first relate the question of the non-negativity of (4.3.3) subject to (4.3.4)-(4.3.6) to that of the non-negativity of a set of linear-quadratic problems.

THEOREM 4.3.1 A sufficient condition for $J[u(\cdot)] \geqslant 0$ for all piecewise continuous $u(\cdot)$ for which (4.3.5) and (4.3.6) hold, is that $\hat{J}[u(\cdot);y(\cdot),z(\cdot)] \geqslant 0$ for all piecewise continuous $u(\cdot)$ and for all $y(\cdot),z(\cdot) \in YZ$ where

$$\hat{J}[u(\cdot);y(\cdot),z(\cdot)] = \int_0^T \{\tfrac{1}{2}x^T Q(y,z,t)x + u^T C(y,z,t)x + \tfrac{1}{2}u^T R(y,z,t)u\}dt + \tfrac{1}{2}x^T(T)Q_T(y(T))x(T) \qquad (4.3.7)$$

and

$$\dot{x} = A(y,z,t)x + B(y,z,t)u \ ; \quad x(0) = 0 \qquad (4.3.8)$$

$$D(y(T))x = 0 \qquad (4.3.9)$$

$$\begin{pmatrix} x \\ u \end{pmatrix} \in \Omega_{xu} \text{ for all } t \in [0,T] \qquad (4.3.10)$$

$$YZ = \{y(\cdot),z(\cdot) \,|\, y(\cdot) \text{ is absolutely continuous, } z(\cdot) \text{ is piecewise continuous on } [0,T]\text{, and } \begin{pmatrix} y \\ z \end{pmatrix} \in \Omega_{xu}\}. \qquad (4.3.11)$$

PROOF Clearly all solutions of (4.3.4) together with their corresponding inputs belong to YZ. It follows, then, that we can set $y(\cdot) = x(\cdot)$ and $z(\cdot) = u(\cdot)$ in (4.3.7)-(4.3.9) so that for this particular choice $J[u(\cdot)] = \hat{J}[u(\cdot);x(\cdot),u(\cdot)] \geqslant 0$.

Theorem 4.3.1 may be stated differently, as follows.

THEOREM 4.3.2 A sufficient condition for $J[u(\cdot)] \geqslant 0$ for all piecewise continuous $u(\cdot)$ for which (4.3.5) and (4.3.6) hold, is that

$$\inf_{y(\cdot),z(\cdot)\, \varepsilon\, YZ} \; \inf_{u(\cdot)} \hat{J}[u(\cdot);y(\cdot),z(\cdot)] \geqslant 0 \quad (4.3.12)$$

subject to (4.3.8)-(4.3.11).

Note that for fixed $y(\cdot)$, $z(\cdot)$ the formulation specified by (4.3.7)-(4.3.10) is identical to that of (4.2.1)-(4.2.4) so that all the sufficiency conditions of Section 4.2 which hold also if Q, C, R, A, B are piecewise continuous, are applicable. The complication that arises is that the conditions must be tested for all the Q, C, R, Q_T, A, B, D realized by (4.3.11). Note further that the class of $y(\cdot)$, $z(\cdot)$ which must be tested can be reduced in size by replacing YZ by the set $\overline{YZ}$ where

$$\overline{YZ} = \{y(\cdot),z(\cdot) \,|\, \dot{y} = A(y,z,t)y+B(y,z,t)z,\ y(0) = 0,$$

$$D(y(T))y(T) = 0, \begin{pmatrix} y \\ z \end{pmatrix} \varepsilon\, \Omega_{xu}\} \quad (4.3.13)$$

which however requires the repeated integration of the non-linear dynamic equation (4.3.4) and the satisfaction of (4.3.5).

4.3.3 An Application

We here demonstrate an application of Theorem 4.3.2 which was suggested by a note of Speyer and Shaked [10]. However their theory appears to contain some questionable steps.

We consider the non-negativity of

$$J[u(\cdot)] = \int_0^T \tfrac{1}{2}(x^T Q x + u^T R u)dt + \tfrac{1}{2}x^T(T)Q_T x(T) \qquad (4.3.14)$$

where

$$\dot{x} = Ax + B(x)u \ , \quad x(0) = 0 \qquad (4.3.15)$$

$B(x) \in R^{n \times m}$ is given by

$$B(x) = \begin{pmatrix} x_{k_1} & & & & 0 \\ & x_{k_2} & & & \\ & & x_{k_3} & & \\ & & & \ddots & \\ 0 & & & & x_{k_m} \\ \hdotsfor{5} \\ & & 0 & & \end{pmatrix}, \; k_i \in \{1,\ldots,n\} \ , \ i=1,\ldots,m \qquad (4.3.16)$$

and where

$$x_{i\,min} \leqslant x_i(t) \leqslant x_{i\,max} \ . \qquad (4.3.17)$$

The matrices Q, R and A are assumed to be functions only of time and

$$R = I > 0 \text{ for all } t \in [0,T]. \qquad (4.3.18)$$

Theorem 4.3.2 states that $J[u(\cdot)]$ is non-negative if

$$\underset{y(\cdot)\in YZ}{\text{Inf}} \ \underset{u(\cdot)}{\text{Inf}} \int_0^T \tfrac{1}{2}(x^T Q x + u^T u)dt + \tfrac{1}{2}x^T(T)Q_T x(T) \geqslant 0 \qquad (4.3.19)$$

subject to

$$\dot{x} = Ax + B(y)u \ , \quad x(0) = 0 \tag{4.3.20}$$

where x(t) satisfies (4.3.17).

If we drop the last-mentioned requirement, which then implies a more restrictive version of the theorem, we have that $J[u(\cdot)] \geqslant 0$ for all piecewise continuous controls $u(\cdot)$ which cause (4.3.17) to hold if

$$\underset{y(\cdot)\,\varepsilon\, YZ}{\text{Inf}} \quad \underset{u(\cdot)}{\text{Inf}} \int_{0}^{T} \tfrac{1}{2}(x^T Q x + u^T u)dt + \tfrac{1}{2}x^T(T)Q_T x(T) \geqslant 0 \tag{4.3.21}$$

subject to

$$\dot{x} = Ax + B(y)u \ , \quad x(0) = 0 \tag{4.3.22}$$

where x and u are unrestricted,

$$YZ = \{y(\cdot) | y(\cdot) \text{ is absolutely continuous,}$$

$$\text{and} \quad x_{i\ min} \leqslant y_i(t) \leqslant x_{i\ max}\} \ . \tag{4.3.23}$$

We then have the following theorem.

THEOREM 4.3.3 A sufficient condition for (4.3.21) to hold, is that there exists for all $t \; \varepsilon \; [0,T]$ a continuously differentiable symmetric matrix function of time $S(\cdot)$ which satisfies

$$-\dot{S} = Q + SA + A^T S - SB(\hat{x})B^T(\hat{x})S \ , \quad S(T) = Q_T \tag{4.3.24}$$

where

$$\hat{x}_{k_i} = \max(|x_{k_i\min}|, |x_{k_i\max}|) \; , \; i=1,\ldots,m. \quad (4.3.25)$$

PROOF By dynamic programming we have that the optimal value function

$$V(x,t) = \min_{\substack{y(\tau) \\ \tau \,\varepsilon\, [t,T]}} \; \min_{\substack{u(\tau) \\ \tau \,\varepsilon\, [t,T]}} \int_t^T \tfrac{1}{2}(x^TQx+u^Tu)d\tau + \tfrac{1}{2}x^T(T)Q_Tx(T)$$

$$(4.3.26)$$

satisfies

$$-\frac{\partial V}{\partial t}(x,t) =$$

$$\min_{x_{min} \leq y(t) \leq x_{max}} \; \min_{u(t)} \; [\tfrac{1}{2}(x^TQx+u^Tu)+V_x(x,t)(Ax+B(y)u)]$$

$$(4.3.27)$$

with

$$V(x(T),T) = \tfrac{1}{2}x^T(T)Q_Tx(T). \quad (4.3.28)$$

Upon minimizing (4.3.27) with respect to u(t) we get

$$u(t) = -\,B^T(y)V_x^T \quad (4.3.29)$$

and this substituted back into (4.3.27) yields

$$-\frac{\partial V}{\partial t}(x,t) =$$

$$\min_{x_{min} \leq y(t) \leq x_{max}} [\tfrac{1}{2}x^TQx+V_x(x,t)Ax-\tfrac{1}{2}V_x(x,t)B(y)B^T(y)V_x^T(x,t)].$$

$$(4.3.30)$$

Now because of the special form of B(y) - see (4.3.16) - we have that the quantity in square brackets in (4.3.30) is minimized by setting

$$y_{k_i} = \hat{x}_{k_i} \triangleq \max(|x_{k_i\min}|,|x_{k_i\max}|), \quad i=1,\ldots,m \tag{4.3.31}$$

irrespective of the value of $V_x(x,t)$. This then yields

$$-\frac{\partial V}{\partial t}(x,t) = \tfrac{1}{2}x^TQx + V_x(x,t)Ax - \tfrac{1}{2}V_x(x,t)B(\hat{x})B^T(\hat{x})V_x^T(x,t) \tag{4.3.32}$$

which is solved by

$$V(x,t) = \tfrac{1}{2}x^TS(t)x\ , \quad t \in [0,T]\ . \tag{4.3.33}$$

Satisfaction of (4.3.21) follows upon noting that V(0,0) = 0.

4.4 Conclusion

In this chapter we first stated the more important theorems for non-negativity of linear-quadratic functionals developed in [1] and elsewhere. Certain of our theorems are novel statements of non-negativity conditions. Section 4.2 was devoted to extensions of these theorems when the state and control vectors are required to belong to general constraint sets. When the constraint sets are specified by linear inequalities, a generalized Riccati equation emerges. In Section 4.3 we formulated a general class of non-linear constrained functionals and showed how the theorems of Section 4.2 are applicable. The main complication that arises in the non-linear case is that the conditions have to be tested for all time functions belonging to a specified set. This seems to be a common complication in the analysis of general non-linear systems and can only be obviated by suitable bounding techniques, an example of which is provided by an application of Theorem 4.3.2 to a non-linear problem of special structure.

In the next chapter we investigate the question of controllability of a linear system when the control vector is required to belong to a general constraint set.

4.5 References

[1] BELL, D.J. & JACOBSON, D.H. Singular Optimal Control Problems. Academic Press, New York and London, 1975.

[2] COPPEL, W.A. Linear-quadratic Optimal Control. Proc. Royal Soc. of Edinburgh, 73A, 1974, pp. 271-288.

[3] MOLINARI, B.P. Non-negativity of a Quadratic Functional. SIAM J. Control, 13, 1975, pp. 792-806.

[4] ANDERSON, B.D.O. Partially Singular Linear-quadratic Control Problems. IEEE Trans. Automatic Control, AC-18, 1973, pp. 407-409.

[5] KRASNER, N.F. Semi-separable Kernels in Linear Estimation and Control. Information Systems Laboratory, Center for Systems Research, Stanford University, California, U.S.A., Technical Report 7001-7, 1974, 319 p.

[6] JACOBSON, D.H. A General Sufficiency Theorem for the Second Variation. J. Math. Anal. Appl., 34, 1971, pp. 578-589.

[7] JACOBSON, D.H. & GETZ, W.M. Remarks on a Paper by V.B. Haas and Partially Singular Extremals. CSIR Special Report, WISK 208, May, 1976.

[8] DAVISON, E.J. & KUNZE, E.G. Some Sufficient Conditions for the Global and Local Controllability of Non-linear Time-varying Systems. SIAM J. Control, 8, 1970, pp. 489-497.

[9] WERNLI, A. & COOK, G. Sub-optimal Control for the Non-linear Quadratic Regulator Problem. Automatica, 11, 1975, pp. 75-84.

[10] SPEYER, J.L. & SHAKED, U. Minimax Design for a Class of Linear Quadratic Problems with Parameter Uncertainty. IEEE Trans. Automatic Control, AC-19, 1974, pp. 158-159.

5. CONTROLLABILITY OF CONSTRAINED LINEAR AUTONOMOUS SYSTEMS

5.1 Introduction

In optimal control and generally in the control of dynamic systems there are certain system-theoretic concepts which play vital roles. One of these is stability, but perhaps the most important is controllability. Loosely speaking we call a dynamic system $\dot{x} = f(x,u)$ controllable if there exists an admissible control function $u(\cdot)$ which transfers the state of the system from an arbitrary initial value x_0 to the origin of the state space in finite time. Intuitively the concept of controllability is basic to the problem of designing minimal time controllers, viz. controllers which steer x_0 to the origin of the state space in minimum time.

This chapter is devoted to a study of the controllability properties of the linear autonomous dynamic system

$$\dot{x} = Ax + Bu \tag{5.1.1}$$

where $A \varepsilon R^{n \times n}$, $B \varepsilon R^{n \times m}$ are constant matrices and $x \varepsilon R^n$, $u \varepsilon R^m$ and where the control variable u is required to satisfy the constraint

$$u(t) \varepsilon \Omega \subseteq R^m, \quad t \varepsilon [0,\infty). \tag{5.1.2}$$

The following sections review certain well-known conditions for controllability of (5.1.1) when $\Omega = R^m$ and when 0 belongs to the interior of Ω ($0 \varepsilon \text{Int}(\Omega)$). The controllability of (5.1.1) when $0 \notin \text{Int}(\Omega)$ is then investigated in detail and new results are presented pertaining to the 'arbitrary-interval null-controllability' of (5.1.1) subject to (5.1.2).

5.1.1 Unconstrained Case

When $\Omega \equiv R^m$, conditions for the controllability of (5.1.1) are much simplified. First we define precisely the concept of controllability.

DEFINITION 5.1.1 The linear autonomous system (5.1.1) is controllable if for each $x_o \in R^n$ there exists a bounded measurable control function which steers x_o to the origin of the state space in finite time.

Necessary and sufficient conditions for controllability were supplied by R.E. Kalman during the early 1960s and are as follows.

THEOREM 5.1.1 A necessary and sufficient condition for (5.1.1) to be controllable when $\Omega \equiv R^m$ is that

$$\text{Rank}(Q) = n \tag{5.1.3}$$

where

$$Q \triangleq [B, AB, \ldots, A^{n-1}B]. \tag{5.1.4}$$

Equivalently,

$$W(0,t) \triangleq \int_0^t \Phi(t,\tau)BB^T\Phi^T(t,\tau)d\tau > 0 \text{ (positive-definite)} \tag{5.1.5}$$

for all $t > 0$ where $\Phi(t,\tau)$ satisfies

$$\frac{d}{dt}\Phi(t,\tau) = A\Phi(t,\tau), \quad \Phi(\tau,\tau) = I \tag{5.1.6}$$

and is the transition matrix associated with $\dot{x} = Ax$.

Inequality (5.1.5) actually allows one to compute a continuous control function which steers x_0 to 0.

THEOREM 5.1.2 The control function

$$u(t) = -B^T\Phi^T(T,t)W^{-1}(0,T)\Phi(T,0)x_0, \quad t \in [0,T] \quad (5.1.7)$$

steers x_0 to the origin in time T.

PROOF The solution of (5.1.1) at time T is given by

$$x(T) = \Phi(T,0)x_0 + \int_0^T \Phi(T,\tau)Bu(\tau)d\tau \ . \quad (5.1.8)$$

Substituting (5.1.7) into (5.1.8) and using (5.1.5) yields

$$x(T) = \Phi(T,0)x_0 - \Phi(T,0)x_0 = 0 \ . \quad (5.1.9)$$

Note that the rank condition (5.1.3) is easy to check, especially when n is small. On the other hand (5.1.3) is not the correct condition when A and B are time-varying, but (5.1.5) remains valid. Note also that controllability here implies not only that there is a control function $u(\cdot)$ which steers x_0 to 0 in finite time but also that there is a control function $u(\cdot)$, given by (5.1.7) which performs this function in an arbitrary time interval $[0,T]$. Thus when $\Omega \equiv R^m$ controllability actually implies 'arbitrary-interval controllability', and the converse is trivially also true.

5.1.2 Zero Interior to Ω

We assume here that $\Omega \subseteq R^m$ is arbitrary but that

$$0 \in \mathrm{Int}(\Omega). \tag{5.1.10}$$

In general, whether or not (5.1.10) is satisfied, it would be too much to expect that (5.1.1) is controllable subject to the restriction (5.1.2). Therefore we introduce the following definition.

DEFINITION 5.1.2 The dynamic system (5.1.1) is null-controllable if there exists an open set V in R^n which contains the origin and for which any $x_0 \in V$ can be controlled to the origin in a finite time by a bounded, measurable control function.

Clearly null-controllability is really controllability in a sphere which surrounds the origin and so controllability implies null-controllability but the converse is not usually true.

We can now state the relevant null-controllability theorem.

THEOREM 5.1.3 System (5.1.1) is null-controllable if (5.1.10) holds, if and only if

$$\mathrm{Rank}(Q) = n. \tag{5.1.11}$$

PROOF If (5.1.1) is null-controllable it is easy to show that (5.1.1) is controllable if the restraint (5.1.2) is removed: condition (5.1.11) then follows. On the other hand, (5.1.11) implies that (5.1.7) is well defined and so for $x_0 \in \{x \mid \|x\| \leqslant \varepsilon_1\}$ it follows that $\|u(t)\| \leqslant \varepsilon_2(\varepsilon_1)$ for $t \in [0,T]$ where $\varepsilon_2(\varepsilon_1) \to 0$ as $\varepsilon_1 \to 0$. As $0 \in \mathrm{Int}(\Omega)$ it is clear that for ε_1 sufficiently small, $u(t) \in \Omega$, $t \in [0,T]$.

Note that as T is arbitrary in the above construction there

exists for each $T > 0$ a set $V(T)$ as specified in Definition (5.1.2). Therefore when $0 \in Int(\Omega)$ null-controllability is, actually, equivalent to 'arbitrary-interval null-controllability'.

5.1.3 Reachable Sets

The notion of a reachable set and certain of its properties play important roles in controllability theory, as we shall see in the following pages. We now define the reachable sets $R_\Omega(t)$, $R_{CH(\Omega)}(t)$ and R_∞. See Section 5.A, the Appendix, for certain set-theoretic definitions.

DEFINITION 5.1.3 The reachable set $R_\Omega(t)(R_{CH(\Omega)}(t))$ at time t is defined to be the set of all points in R^n to which the origin can be steered at time t by a bounded measurable control function $u(\cdot)$ which satisfies $u(\tau) \in \Omega$ ($u(\tau) \in Convhull(\Omega)$) for all $\tau \in [0,t]$. The reachable set R_∞ is the union over positive t of the sets $R_\Omega(t)$.

THEOREM 5.1.4 [1] Consider the linear process (5.1.1) with constraint set $\Omega \subseteq R^m$. The reachable set $R_\Omega(t)$ at time t is convex. If in addition Ω is compact then $R_\Omega(t)$ is convex, compact, and varies continuously with t on $t \geqslant 0$ and $R_{CH(\Omega)}(t) = R_\Omega(t)$.

PROOF The first part of the theorem and the compactness of $R_\Omega(t)$ if Ω is compact follow direct from Lemma 4A [1, p. 163] on the range of a vector measure. The equivalence of $R_\Omega(t)$, $R_{CH(\Omega)}(t)$ is proved in Theorem 1A [1, p. 164] by noting that $R_\Omega(t) \subseteq R_{CH(\Omega)}(t)$ and that these sets are convex and compact if Ω is compact and by proving that $R_\Omega(t)$ is dense in $R_{CH(\Omega)}(t)$.

The next theorem shows that the nature of the reachable sets of (5.1.1) subject to (5.1.2) is directly related to the controllability of the negative of system (5.1.1).

THEOREM 5.1.5 The reachable set R_∞ of system (5.1.1) subject to (5.1.2) contains a neighbourhood of the origin if and only if the system

$$\dot{x} = -Ax - Bu \tag{5.1.12}$$

subject to (5.1.2) is null-controllable.

PROOF Suppose that $u(t)$, $t \in [0,T]$ steers (5.1.12) from x_0 to 0 at $t = T$. Then it is easy to show that $u(T-t)$, $t \in [0,T]$ steers (5.1.1) from 0 to x_0 at $t = T$. The converse is also easily verified.

The next theorem is of fundamental importance in studies of null-controllability.

THEOREM 5.1.6 [2] Suppose that there exists a vector $u \in \Omega$ such that $Bu = 0$. Then, the origin is interior to the reachable set R_∞ of (5.1.1) if and only if there is no non-zero $v \in R^n$ such that

$$v^T e^{At} Bu \leqslant 0 \text{ for all } t > 0 \text{ and for all } u \in \Omega . \tag{5.1.13}$$

PROOF By Theorem 5.1.4, $R_\Omega(t)$ is convex for all $t > 0$. Furthermore the fact that there is a $u \in \Omega$ such that $Bu = 0$ implies that $R_\Omega(t_1) \subseteq R_\Omega(t_2)$ for $t_1 \leqslant t_2$ and $0 \in R_\Omega(t)$ for all $t > 0$. Therefore it follows that R_∞ is the union of nested increasing convex sets which contain the origin, and is thus a convex set which contains the origin. If in fact the origin is interior to R_∞ there exist n + 1 points in R_∞ whose convex hull contains the origin as an interior point. It follows that these n + 1 points must be contained in $R_\Omega(t)$ for some $t > 0$ since these sets are increasing. Therefore it is clear that the origin is interior to R_∞ if and only if the origin is interior to $R_\Omega(t)$ for some $t > 0$. Consequently if the origin is not interior to R_∞ there exists a $v \varepsilon R^n$ such that

$$v^T x(t) \leqslant 0 \qquad (5.1.14)$$

for all $t > 0$ and all admissible controls $u(\cdot)$. Inequality (5.1.14) is just

$$v^T \int_0^t e^{A(t-\tau)} Bu(\tau)d\tau \leqslant 0 \qquad (5.1.15)$$

and it follows by continuity and a special choice of $u(\cdot)$ that

$$v^T e^{At} Bu \leqslant 0 \text{ for all } t > 0 \text{ and for all } u \varepsilon \Omega. \qquad (5.1.16)$$

On the other hand, if the origin is interior to R_∞ there cannot exist a vector $v \varepsilon R^n$ which satisfies (5.1.16), as the existence of such a vector would imply inequality (5.1.14).

5.1.4 Zero Interior to Convhull(Ω)

The last theorem and Theorems 5.1.4 and 5.1.5 may now be combined to yield the following result which is similar to Theorem 5.1.3.

THEOREM 5.1.7 Suppose that Ω is compact and that $0 \in \text{Int}(\text{Convhull}(\Omega))$. Then, system (5.1.1) subject to (5.1.2) is null-controllable if and only if

$$\text{Rank}(Q) = n. \tag{5.1.17}$$

PROOF As Ω is compact, Theorem 5.1.4 implies that Ω can be replaced by $\text{Convhull}(\Omega)$ as far as reachable sets are concerned. As zero belongs to $\text{Convhull}(\Omega)$ we have that there exists a vector $u \in \text{Convhull}(\Omega)$ such that $Bu = 0$. Theorem 5.1.6 then implies that zero is interior to R_∞ if and only if there is no non-zero $v \in R^n$ such that

$$v^T e^{At} Bu \leqslant 0 \text{ for all } t > 0 \text{ and for all } u \in \text{Convhull}(\Omega). \tag{5.1.18}$$

We now prove that this is true if and only if (5.1.17) holds. Suppose first that Q does not have rank n. Then, there exists a vector $v \in R^n$ such that

$$v^T Q = 0. \tag{5.1.19}$$

Now it is well known that this implies that

$$v^T e^{At} B = 0 \tag{5.1.20}$$

so that

$$v^T e^{At} Bu = 0 \text{ for all } t > 0 \text{ and all } u \in \text{Convhull}(\Omega) \tag{5.1.21}$$

which contradicts the statement that zero is interior to R_∞.

Suppose now that Q has rank n but that $v^T e^{At} Bu = 0$ for all t and all $u \varepsilon$ Convhull(Ω). Then by setting $t = 0$ we obtain

$$v^T Bu = 0 \text{ for all } u \ \varepsilon \ \text{Convhull}(\Omega) \tag{5.1.22}$$

and successive differentiations of $v^T e^{At} Bu$ and evaluations at $t = 0$ yield

$$v^T A^i Bu = 0 \text{ for all } u \ \varepsilon \ \text{Convhull}(\Omega), \quad i=1,\ldots,n-1. \tag{5.1.23}$$

As zero is interior to Convhull(Ω), Equations (5.1.22) and (5.1.23) imply that Q has rank less than n, which is a contradiction. Therefore (5.1.17) implies that there exists $u \varepsilon$ Convhull(Ω) such that $v^T e^{At} Bu$ is not identically zero for all t. Now either $v^T e^{At} Bu$ changes sign as a function of t for this fixed $u \varepsilon$ Convhull(Ω) or it is of one sign. Suppose in fact that $v^T e^{At} Bu \leqslant 0$ (not identically zero, as we have proved). As zero is interior to Convhull(Ω) we can switch the sign of $v^T e^{At} Bu$ to be positive (whenever it is non-zero) by reversing the sign of u so that (5.1.18) does not hold for any $v \ \varepsilon \ R^n$. This completes the proof that zero is interior to R_∞ if and only if (5.1.17) holds. The observation that

$$\text{Rank}[B,AB,\ldots,A^{n-1}B] = \text{Rank}[-B,AB,-A^2B,\ldots,(-1)^n A^{n-1}B] \tag{5.1.24}$$

then yields the theorem.

5.1.5 Global and Non-linear Results

As remarked prior to Definition 5.1.2 controllability (as distinct from null-controllability) cannot in general be expected of (5.1.1) when it is subjected to the constraint (5.1.2). However, the following theorems provide conditions which guarantee that controllability (null-controllability in the large) can be achieved by (5.1.1) subject to (5.1.2).

THEOREM 5.1.8 [1] Consider the system (5.1.1) subject to the control constraint (5.1.2). Suppose that zero belongs to the interior of Ω, that Rank(Q) = n, and that A is stable ($\mathrm{Re}(\lambda) < 0$). Then (5.1.1) is controllable.

PROOF The stability of $\dot{x} = Ax$ ensures that any initial point $x_0 \in R^n$ can be steered by $u(\cdot) \equiv 0$ until x(t) approaches 0 and therefore enters the domain of null-controllability of (5.1.1). But then x(t) can be steered to the origin in a finite time.

When the control function is scalar the strict inequality on $\mathrm{Re}(\lambda)$ can be relaxed, but then compactness of Ω is assumed.

THEOREM 5.1.9 [1] Suppose that m = 1 and that Ω is a compact set which contains zero in its interior. Then (5.1.1) is controllable if and only if Rank(Q) = n and every eigenvalue λ of A satisfies $\mathrm{Re}(\lambda) \leqslant 0$.

When $m \geqslant 1$ we have the following theorem.

THEOREM 5.1.10 [1] Suppose that Ω is a compact set which contains zero in its interior. Assume that no two Jordan canonical blocks of A contain equal eigenvalues of A. Then (5.1.1) is controllable if and only if Rank(Q) = n and every eigenvalue λ of A satisfies $Re(\lambda) \leqslant 0$.

Finally, we remark that null-controllability of a non-linear system can sometimes be deduced from the null-controllability of its linearization. This is made precise in the following theorem [1].

THEOREM 5.1.11 Consider the non-linear autonomous system

$$\dot{x} = f(x,u) \tag{5.1.25}$$

where $f:R^{n+m} \to R^n$ is once continuously differentiable in x and u. Suppose that zero is interior to the constraint set $\Omega \subseteq R^m$, and assume that

$$f(0,0) = 0 \tag{5.1.26}$$

$$\text{Rank}\,[B,AB,\ldots,A^{n-1}B] = n \tag{5.1.27}$$

where

$$A = \frac{\partial f}{\partial x}(0,0), \quad B = \frac{\partial f}{\partial u}(0,0) \; . \tag{5.1.28}$$

Then (5.1.25) is null-controllable.

5.2 Zero Not Interior to Constraint Set or its Convex Hull

5.2.1 Oscillatory Systems

In this section we devote attention to recent work on the controllability of (5.1.1) subject to (5.1.2) when $0 \notin Int(\Omega)$. Initially research on this problem was motivated by the question [3]: can the motion of a simple pendulum be brought to rest in a finite time by the application of a unit force acting only in one direction? In mathematical terms one asks whether the system

$$\dot{x}_1 = x_2 \qquad\qquad (5.2.1)$$
$$\dot{x}_2 = -x_1 + u$$

is (null-) controllable when

$$u(t) \in [0,1], \quad t \in [0,\infty). \qquad (5.2.2)$$

Clearly (5.2.1) is controllable when $u(t) \in R^1$ and is not restricted according to (5.2.2). Furthermore if $u(t) \in [-\varepsilon,1]$, $\varepsilon > 0$ Theorem 5.1.3 guarantees that (5.2.1) is null-controllable.

Saperstone and Yorke were the first to prove the following result when $\Omega = [0,1]$.

THEOREM 5.2.1 Suppose that $m = 1$ and $\Omega = [0,1]$. The system (5.1.1) is null-controllable if and only if all the eigenvalues of A have non-zero imaginary parts and $\text{Rank}(Q) = n$.

This interesting theorem implies that if n is odd (implying at least one real eigenvalue) (5.1.1) is not null-controllable. Note that the conditions of the theorem are satisfied by (5.2.1) and (5.2.2) so that (5.2.1) is indeed null-controllable subject to the constraint that $u(t) \in [0,1]$. Note also that Theorem 5.1.4 allows one to replace the interval $[0,1]$ in Theorem 5.2.1 by the set $\{0,1\}$ consisting of the two end points of the compact interval.

It turns out that Theorem 5.2.1 is not the best that can be obtained when $0 \notin \text{Int}(\Omega)$ and $0 \notin \text{Int}(\text{Convhull}(\Omega))$. The appropriate generalization of Theorem 5.2.1 which we investigate fully in the next section is due to Brammer [2].

5.2.2 Brammer's Theorem

The following theorem is a major generalization of Theorem 5.2.1 though it is simple in form. However, Brammer's proof is exceedingly lengthy and intricate so that we prefer to present Hãjek's short proof [4].

THEOREM 5.2.2 [2] Suppose that

$$\text{there is a } u \in \Omega \text{ such that } Bu = 0 \tag{5.2.3}$$

and that

$$\text{the set Convhull}(\Omega) \text{ has non-empty interior in } R^m. \tag{5.2.4}$$

Then (5.1.1) subject to (5.1.2) is null-controllable if and only if

$$\text{Rank}(Q) = n \tag{5.2.5}$$

and

there is no real eigenvector v of A^T satisfying

$$v^T Bu \leqslant 0 \text{ for all } u \in \Omega. \tag{5.2.6}$$

Clearly (5.2.6) is trivially satisfied if A has no real eigenvalues so that Theorem 5.2.2 reduces to Theorem 5.2.1 when $m = 1$ and $\Omega = [0,1]$.

Hãjek's proof of the above theorem requires the following lemmas, which we only state.

LEMMA 5.2.1 Suppose that $A \in R^{n \times n}$ and V is a closed convex cone in R^n such that v and -v belong to V only if v = 0. If V is invariant under e^{At}, viz. $e^{At}V \subseteq V$ for all $t \geqslant 0$ then V contains an eigenvector of A.

LEMMA 5.2.2 The linear hull of $\{e^{At}Bu | t \geqslant 0, u \in \Omega\}$ is R^n if and only if Rank(Q) = n (provided that (5.2.3), (5.2.4) hold).

LEMMA 5.2.3 The system $\dot{x} = Ax + Bu$ satisfies (5.2.5), (5.2.6) if and only if the negative system $\dot{x} = -Ax - Bu$ does.

PROOF OF THEOREM 5.2.2 As a consequence of Theorems 5.1.5, 5.1.6 and Lemma 5.2.3 we have that (5.1.1) is null-controllable subject to conditions (5.2.3) and (5.2.4) if and only if there is no non-zero vector $v \in R^n$ such that

$$v^T e^{At} Bu \leqslant 0 \text{ for all } t > 0 \text{ and for all } u \in \Omega. \tag{5.2.7}$$

If (5.2.5) fails it is easy to show via Lemma 5.2.2 that there is a non-zero vector $v \in R^n$ such that

$$v^T e^{At} Bu = 0 \text{ for all } t \text{ and all } u \in \Omega. \tag{5.2.8}$$

If (5.2.6) is satisfied by a real eigenvector $v \in R^n$ of A^T we have that

$$v^T e^{At} Bu = v^T Bu . e^{\lambda t} \tag{5.2.9}$$

where λ is the real eigenvalue associated with v. It follows from (5.2.6) and (5.2.9) that

$$v^T e^{At} Bu \leqslant 0 \text{ for all } t > 0 \text{ and all } u \in \Omega. \qquad (5.2.10)$$

We conclude from the above that the conditions (5.2.5) and (5.2.6) are necessary.

Next we assume that (5.1.1) is not null-controllable but that the conditions (5.2.5), (5.2.6) hold; this leads to a contradiction.

Let W be the set of all linear combinations with non-negative coefficients of points $e^{At}Bu$ for $t > 0$, $u \in \Omega$. Then from (5.2.7) we have that if (5.1.1) is not null-controllable there is a $v \in R^n$ such that

$$v^T w \leqslant 0 \text{ for all } w \in W. \qquad (5.2.11)$$

Now W is a closed convex cone in R^n which by Lemma 5.2.2 and condition (5.2.5) satisfies

$$W + (-W) = R^n \qquad (5.2.12)$$

and is obviously invariant under e^{At}, viz.

$$e^{At} W \subseteq W \text{ for all } t \geqslant 0. \qquad (5.2.13)$$

Next we note that the set V consisting of those vectors $v \in R^n$ which satisfy

$$v^T w \leqslant 0 \text{ for each } w \in W \qquad (5.2.14)$$

is also a closed convex cone in R^n. Furthermore (5.2.12) guarantees that v and -v both belong to V only if v = 0. We see then that the invariance of W under e^{At} implies invariance of V under $e^{A^T t}$. Lemma 5.2.1 then implies that V contains an eigenvector v of A^T. But then (5.2.6) is satisfied by an eigenvector of A^T, which contradicts our assumption, and the sufficiency of the conditions of the theorem is proved.

We note that Heymann and Stern [5] have also offered a geometric proof of the above theorem. However, Hājek's approach seems to us to be the simplest.

5.2.3 Global Results

The following theorem shows that if Ω is suitably unbounded, the reachable set R_∞ of (5.1.1) is all of R^n.

THEOREM 5.2.3 [2] If Ω is a cone with vertex at the origin and has non-empty interior in R^m then (5.2.5), (5.2.6) are necessary and sufficient for the reachable set R_∞ of (5.1.1) to be all of R^n.

PROOF The necessity of (5.2.5) and (5.2.6) is already proved. For sufficiency we first note that as Ω is a cone with vertex at the origin so also is the reachable set R_∞. Furthermore the reachable set is convex and contains the origin in its interior if (5.2.5), (5.2.6) are satisfied. Combining these two properties of R_∞ we obtain the desired result.

COROLLARY 5.2.1 If Ω is a cone with vertex at the origin and has non-empty interior in R^m then (5.2.5), (5.2.6) are necessary and sufficient for (5.1.1) to be controllable.

PROOF Theorem 5.2.3 implies via the proof of Theorem 5.1.5 that (5.1.12) is controllable if and only if R_∞ is all of R^n. It then follows under the stated assumptions that (5.1.12) is controllable if and only if (5.2.5) and (5.2.6) are satisfied. However (5.1.1) satisfies (5.2.5), (5.2.6) if and only if (5.1.12) does, and the corollary is proved.

As Theorem 5.1.8 indicates it is not always necessary that Ω be unbounded for R_∞ to be all of R^n or for (5.1.1) to be controllable. The following theorem is the appropriate generalization of Theorem 5.1.8 and the proof is the same.

THEOREM 5.2.4 Suppose that the conditions of Theorem 5.2.2 are satisfied and that A is stable ($Re(\lambda) < 0$). Then (5.1.1) is controllable.

5.3 Arbitrary-Interval Null-Controllability

As noted in Section 5.1.1, the necessary and sufficient conditions for controllability of (5.1.1) subject to (5.1.2) when $\Omega = R^m$ are such that if (5.1.1) is controllable then any point $x_0 \in R^n$ can be transferred to the origin of the state space in time $T > 0$, where T is arbitrary. This is easily seen to be true as a consequence of (5.1.7) which explicitly constructs a control function $u(t)$, $0 \leq t \leq T$ which transfers x_0 at time 0 to 0 at time T.

When $\Omega \subset R^m$ and $0 \in Int(\Omega)$ we saw that (5.1.1) is null-controllable if and only if $Rank(Q) = n$. In fact, the proof of Theorem 5.1.3 shows that if (5.1.1) is null-controllable when $0 \in Int(\Omega)$ then there is a neighbourhood $V(T)$ of the origin which can be transferred to the origin in time $T > 0$, where T is arbitrary.

On the other hand if $0 \notin Int(\Omega)$ we are not always able to transfer a neighbourhood of the origin, whose size depends upon T, to the origin in time $T > 0$. The example defined by (5.2.1) and (5.2.2) is a case in point. Indeed if $x_2(0) > 0$ application of the non-negative control $u(t) \in [0,1]$ only increases the value of $x_2(t)$ away from its starting value of $x_2(0)$. More precisely, we have that

$$x_1(t) = x_1(0)\cos(t) + x_2(0)\sin(t) + \int_0^t \sin(t-\tau)u(\tau)d\tau \tag{5.3.1}$$

and

$$x_2(t) = - x_1(0)\sin(t) + x_2(0)\cos(t) + \int_0^t \cos(t-\tau)u(\tau)d\tau \tag{5.3.2}$$

so that when $x_1(0) = 0$ we have

$$x_2(t) = x_2(0)\cos(t) + \int_0^t \cos(t-\tau)u(\tau)d\tau \tag{5.3.3}$$

which is positive for $x_2(0) > 0$ and $t < \frac{\pi}{2}$ implying that $x_2(t)$ cannot be steered to zero from a positive starting value in the time $T < \frac{\pi}{2}$. Note, however, that the system (5.2.1) subject to (5.2.2) <u>is</u> null-controllable.

The above observations therefore lead us to a study of 'arbitrary-interval null-controllability' which we define precisely in the next section. A number of the results presented in the following sections were developed in collaboration with M. Pachter [6].

5.3.1 Preliminary Results

DEFINITION 5.3.1 We call (5.1.1) arbitrary-interval null-controllable if for each $T > 0$ there exists an open set $V(T)$ in R^n which contains the origin and for which any $x_0 \in V(T)$ can be controlled to the origin at time T by a bounded measurable control function.

The next theorem relates the properties of the reachable sets of (5.1.1) to arbitrary-interval null-controllability.

THEOREM 5.3.1 The system (5.1.1) subject to (5.1.2) is arbitrary-interval null-controllable if and only if for each $T > 0$ the reachable set $R_\Omega(T)$ of (5.1.1) contains the origin in its interior.

PROOF For fixed but arbitrary T we have

$$R_\Omega(T) = \{\int_0^T e^{A(T-\tau)}Bu(\tau)d\tau \,|\, u(\tau) \in \Omega,\ 0 \leqslant \tau \leqslant T\}. \quad (5.3.4)$$

If (5.1.1) is arbitrary-interval null-controllable we have for each $T > 0$ that

$$V(T) = \{-e^{-AT} \int_0^T e^{A(T-\tau)}Bu(\tau)d\tau \,|\, u(\tau) \in \Omega,\ 0 \leqslant \tau \leqslant T\}. \quad (5.3.5)$$

It follows from (5.3.4), (5.3.5) and the invertibility of e^{AT} that $R_\Omega(T)$ contains the origin in its interior because $V(T)$ does. The converse, namely that $V(T)$ exists if $R_\Omega(T)$ contains the origin in its interior, follows in the same way via the invertibility of e^{AT}.

Note that the relationship between arbitrary-intervul null-controllability and reachability is simpler than that between null-controllability and reachability (cf. Theorems 5.1.5, 5.1.7).

The following theorem exploits the relationship.

THEOREM 5.3.2 A necessary condition for arbitrary-interval null-controllability of (5.1.1) subject to (5.1.2) is that there should exist for each non-zero $v \in R^n$ and for each $T > 0$ a time τ, $0 \leqslant \tau \leqslant T$ and a $u \in \Omega$ such that

$$v^T e^{A\tau} Bu > 0. \tag{5.3.6}$$

PROOF Suppose that (5.1.1) subject to (5.1.2) is arbitrary-interval null-controllable. Theorem 5.3.1 then implies that for each $T > 0$, $R_\Omega(T)$ contains the origin in its interior. This implies that there is no non-zero $v \in R^n$ such that

$$v^T \int_0^T e^{A(T-\tau)} Bu(\tau)d\tau \leqslant 0 \text{ for all } u(\cdot) \text{ such that } u(\tau) \in \Omega,\ 0 \leqslant \tau \leqslant T. \tag{5.3.7}$$

Now if (5.3.6) were not satisfied for each non-zero $v \in R^n$ for some τ, $0 \leqslant \tau \leqslant T$, there would exist a non-zero $v \in R^n$ such that

$$v^T e^{A\tau} Bu \leqslant 0 \text{ for all } \tau,\ 0 \leqslant \tau \leqslant T \text{ and all } u \in \Omega \tag{5.3.8}$$

but this would imply that

$$v^T \int_0^T e^{A(T-\tau)} Bu(\tau)d\tau \leqslant 0 \text{ for all } u(\cdot) \text{ such that } u(\tau) \in \Omega,\ 0 \leqslant \tau \leqslant T \tag{5.3.9}$$

which contradicts the fact that $R_\Omega(T)$ contains the origin in its interior. Therefore the theorem is proved.

We can now deduce an important necessary condition when Ω is bounded.

THEOREM 5.3.3 Assume that Ω is bounded. Then a necessary condition for arbitrary-interval null-controllability of (5.1.1) subject to (5.1.2) is that

$$0 \in \mathrm{Cl}(\mathrm{Convhull}(B\Omega)). \tag{5.3.10}$$

If Ω is compact, (5.3.10) becomes

$$0 \in B\mathrm{Convhull}(\Omega). \tag{5.3.11}$$

PROOF Clearly if the system (5.1.1) subject to (5.1.2) is arbitrary-interval null-controllable, so also is the system

$$\dot{x} = Ax + \tilde{u} \tag{5.3.12}$$

where

$$\tilde{u}(t) \in \mathrm{Cl}(\mathrm{Convhull}(B\Omega)). \tag{5.3.13}$$

Now assume that (5.3.10) does not hold so that there exists a hyperplane strictly separating 0 and $\mathrm{Cl}(\mathrm{Convhull}(B\Omega))$, viz. there is a non-zero $v \in R^n$ such that

$$v^T z < 0 \text{ for all } z \in \mathrm{Cl}(\mathrm{Convhull}(B\Omega)). \tag{5.3.14}$$

However from Theorem 5.3.2 arbitrary-interval null-controllability implies that given a sequence T_n, $T_n > 0$, $T_n \to 0$, there is a sequence τ_n, $0 \leq \tau_n \leq T_n$ and a sequence

$z_n \in \mathrm{Cl}(\mathrm{Convhull}(B\Omega))$ such that $v^T e^{A\tau_n} z_n > 0$. Now because Ω is bounded $\mathrm{Cl}(\mathrm{Convhull}(B\Omega))$ is compact so that there is a sub-sequence $\{z_k\}$ of $\{z_n\}$ which converges to $\bar{z} \in \mathrm{Cl}(\mathrm{Convhull}(B\Omega))$ as $k \to \infty$ and $\tau_k \to 0$. Hence $\lim_{k \to \infty} v^T e^{A\tau_k} z_k = v^T \bar{z} \geqslant 0$, which contradicts (5.3.14). Now Ω compact implies that $\mathrm{Convhull}(\Omega)$ is compact and (5.3.11) follows from (5.3.10).

The following necessary and sufficient condition is the counterpart of Theorem 5.1.6.

THEOREM 5.3.4 Suppose that there exists a vector $u \in \Omega$ such that $Bu = 0$. Then, (5.1.1) subject to (5.1.2) is arbitrary-interval null-controllable if and only if there exists for each non-zero $v \in R^n$ and for each $T > 0$ a time τ, $0 \leqslant \tau \leqslant T$ and a $u \in \Omega$ such that $v^T e^{A\tau} Bu > 0$. Equivalently, there should exist no $v \in R^n$ such that for some $T > 0$, $v^T e^{A\tau} Bu \leqslant 0$, $0 \leqslant \tau \leqslant T$ for all $u \in \Omega$.

PROOF The assumption that there is a $u \in \Omega$ such that $Bu = 0$ implies that $R_\Omega(t_1) \subseteq R_\Omega(t_2)$, $t_1 \leqslant t_2 \leqslant T$. Furthermore $R_\Omega(t)$ is convex (Theorem 5.1.4) and $0 \in R_\Omega(t)$, $t \in (0,T]$, $T > 0$ and arbitrary.

We already showed (Theorem 5.3.2) that the condition of the theorem is necessary for arbitrary-interval null-controllability. To prove sufficiency we show that if the system is not arbitrary-interval null-controllable then the condition of the theorem is violated. If the system is not arbitrary-interval null-controllable then $0 \notin \mathrm{Int}(R_\Omega(T))$ for some $T > 0$ (Theorem 5.3.1). As a consequence of the convexity and the nesting of the reachable sets we see that this implies the existence of a non-zero $v \in R^n$ such that

$$v^T \int_0^t e^{A(t-\tau)} Bu(\tau)d\tau \leqslant 0, \; 0 \leqslant t \leqslant T, \quad \text{for all } u(\cdot) \text{ such that } u(\tau) \in \Omega . \tag{5.3.15}$$

For a particular choice of $u(\cdot)$ this implies that

$$v^T e^{At} Bu \leqslant 0, \; 0 \leqslant t \leqslant T, \text{ for all } u \in \Omega \tag{5.3.16}$$

which establishes the desired contradiction.

5.3.2 Necessary and Sufficient Conditions

In the following pages we shall often refer to (5.1.1) subject to (5.1.2) as the system (A, B, Ω).

We shall assume throughout that

$$\text{there is a } u \in \Omega \text{ such that } Bu = 0. \tag{5.3.17}$$

The next theorem, due to M. Pachter, relates the arbitrary-interval null-controllability of (A, B, Ω) to that of $(A, B, \mathrm{Cl}(\mathrm{Conichull}(\mathrm{Convhull}(\Omega))))$. The uninitiated reader is referred to Section 5.A, the Appendix, for the definitions of Conichull, Convhull and other set-theoretic concepts.

THEOREM 5.3.5 The system (A, B, Ω) is arbitrary-interval null-controllable if and only if the system $(A, B, \mathrm{Cl}(\mathrm{Conichull}(\mathrm{Convhull}(\Omega))))$ is arbitrary-interval null-controllable.

PROOF Theorem 5.3.4 states that (A, B, Ω) is arbitrary-interval null-controllable if and only if for each non-zero $v \in R^n$ and $T > 0$ there is a τ, $0 \leqslant \tau \leqslant T$ and a $u \in \Omega$ such that $v^T e^{A\tau} Bu > 0$. Now, trivially, $v^T e^{A\tau} Bu = u^T B^T e^{A^T\tau} v$ so that the preceding statement is equivalent to the statement that for each non-zero $v \in R^n$ and $T > 0$ there is a τ, $0 \leqslant \tau \leqslant T$ such that $B^T e^{A^T\tau} v \notin \Omega'$, where (see Section 5.A) Ω' is the polar of Ω. Now Ω' is a closed, convex cone so that the separating hyperplane theorem implies that $B^T e^{A^T\tau} v \notin \Omega'$ is equivalent to the existence of a non-zero $a \in R^m$ such that $a^T \omega' \leqslant 0$ for all $\omega' \in \Omega'$ and $a^T B^T e^{A^T\tau} v > 0$. It follows from this that (A, B, Ω) is arbitrary-interval null-controllable if and only if for each non-zero $v \in R^n$ and $T > 0$ there is a τ, $0 \leqslant \tau \leqslant T$ and $a \in \Omega''$ such that $a^T B^T e^{A^T\tau} v = v^T e^{A\tau} Ba > 0$ where $\Omega'' \triangleq (\Omega')'$. Thus (A, B, Ω) is arbitrary-interval null-controllable if and only if (A, B, Ω'') is arbitrary-interval null-controllable.

Now it is clear from the definition of the polar of a set that $\Omega \subseteq \Omega''$ and Ω'' is a closed, convex cone. Therefore it follows that $\text{Convhull}(\Omega) \subseteq \Omega''$, $\text{Conichull}(\text{Convhull}(\Omega)) \subseteq \Omega''$ and $\text{Cl}(\text{Conichull}(\text{Convhull}(\Omega))) \subseteq \Omega''$. Hence $\Omega \subseteq \text{Cl}(\text{Conichull}(\text{Convhull}(\Omega))) \subseteq \Omega''$ so that arbitrary-interval null-controllability with the constraint set $\text{Cl}(\text{Conichull}(\text{Convhull}(\Omega)))$ implies the same with Ω'' and hence with Ω, and the theorem is proved.

The above theorem and its proof indicate that one may work with the 'nicer' closed, convex, cone $\text{Cl}(\text{Conichull}(\text{Convhull}(\Omega)))$

which is equivalent to Cl(Convhull(Conichull(Ω))) for any $\Omega \subseteq R^m$.

The following theorem shows that if (A, B, Ω) is arbitrary-interval null-controllable so is (A, B, $\tilde{\Omega}$) where $\tilde{\Omega}$ is a suitable, bounded, constraint set.

THEOREM 5.3.6 Let $K = \{z \in R^m | \|z\| \leqslant 1\}$.

Then,

(i) the system (A, B, Ω) is arbitrary-interval null-controllable if and only if the system (A, B, Conichull(Ω) $\cap$ K) is arbitrary-interval null-controllable;

(ii) if $0 \in$ Convhull(Ω) then the system (A, B, Ω) is arbitrary-interval null-controllable if and only if the system (A, B, Convhull(Ω) $\cap$ K) is arbitrary-interval null-controllable.

PROOF (i) If we can show that

$$\text{Conichull}(\Omega) = \text{Conichull}(\text{Conichull}(\Omega) \cap K) \quad (5.3.17)$$

then the theorem follows from Theorem 5.3.5. Indeed (5.3.17) would allow us to write

$$\begin{aligned} &\text{Cl}(\text{Conichull}(\text{Convhull}(\text{Conichull}(\Omega) \cap K))) \\ &= \text{Cl}(\text{Convhull}(\text{Conichull}(\text{Conichull}(\Omega) \cap K))) \\ &= \text{Cl}(\text{Convhull}(\text{Conichull}(\Omega))) = \text{Cl}(\text{Conichull}(\text{Convhull}(\Omega))) \end{aligned} \quad (5.3.18)$$

and an application of Theorem 5.3.5 would yield the desired result. We thus proceed now to prove that (5.3.17) holds. First, it is obvious that

$$\text{Conichull}(\Omega) \cap K \subseteq \text{Conichull}(\Omega) \qquad (5.3.19)$$

so that Conichull(Conichull(Ω) $\cap$ K) $\subseteq$ Conichull(Ω).

Next, suppose that $z \in$ Conichull(Ω), $z \neq 0$. Then $y \triangleq \frac{1}{\|z\|} z \in$ Conichull(Ω) so that $y \in$ Conichull(Ω) $\cap$ K. It follows then that $\|z\|.y \in$ Conichull(Conichull(Ω) $\cap$ K) so that $z \in$ Conichull(Conichull(Ω) $\cap$ K), and

$$\text{Conichull}(\Omega) \subseteq \text{Conichull}(\text{Conichull}(\Omega) \cap K). \qquad (5.3.20)$$

The inclusions (5.3.18) and (5.3.19) yield (5.3.17) and (i) is proved.

(ii) Here Theorem 5.3.5 would be again applicable if one could show that

$$\text{Conichull}(\text{Convhull}(\Omega)) = \text{Conichull}(\text{Convhull}(\Omega) \cap K). \qquad (5.3.21)$$

Clearly we have that Convhull(Ω) $\supseteq$ Convhull(Ω) $\cap$ K so that

$$\text{Conichull}(\text{Convhull}(\Omega) \cap K) \subseteq \text{Conichull}(\text{Convhull}(\Omega)). \qquad (5.3.22)$$

Now let $z \in$ Conichull(Convhull(Ω)), $z \neq 0$. Then there is a $\lambda > 0$ such that $\lambda z \in$ Convhull(Ω) and we have that if $\|\lambda z\| \leqslant 1$ then $\lambda z \in$ Convhull(Ω) $\cap$ K so that $z \in$ Conichull(Convhull(Ω) $\cap$ K). If it turns out that $\|\lambda z\| > 1$ then since $0 < \frac{1}{\|\lambda z\|} < 1$ and $0 \in$ Convhull(Ω) we have that $\frac{1}{\|\lambda z\|}\lambda z \in \text{Convhull}(\Omega)$ and $\frac{1}{\|\lambda z\|}\lambda z \in \text{Convhull}(\Omega) \cap K$. Hence λz always belongs to Conichull(Convhull(Ω) $\cap$ K) so that

$$\text{Conichull}(\text{Convhull}(\Omega)) \subseteq \text{Conichull}(\text{Convhull}(\Omega) \cap K). \tag{5.3.23}$$

Inclusions (5.3.22) and (5.3.23) yield (5.3.21) and (ii) is proved.

5.3.3 Necessary Conditions

In this section we provide a necessary condition for arbitrary-interval null-controllability which has an appealing geometric character. First we need the following lemma.

LEMMA 5.3.1 Let C be a closed cone in R^n and let $v \in R^n$ be such that

$$\|v\| = 1,\ v^T c < 0, \text{ for all } c \in C,\ c \neq 0. \tag{5.3.24}$$

Then the set

$$\Delta \triangleq \{x \in R^n | v^T x = -1,\ x \in C\} \tag{5.3.25}$$

is compact.

PROOF As Δ is the intersection of two closed sets it is closed. We now prove that Δ is bounded. We have that $x \in \Delta$ implies that $x = -v + N(v^T)$, where $N(\cdot)$ denotes 'the null space of'. Let us assume that Δ is unbounded, which implies that there is a sequence $x_k \in N(v^T)$ such that $-v + x_k \in C$ and $\|x_k\| \to \infty$ as $k \to \infty$. Clearly the sequence

$y_k \triangleq \frac{x_k}{\|x_k\|} \subseteq N(v^T) \cap \{x | \|x\| \leqslant 1\}$ which is a compact set, so that $y_k \to \bar{y} \in N(v^T)$, viz. $v^T\bar{y} = 0$. Now since C is a cone we have that $z_k = \frac{-v}{\|x_k\|} + \frac{x_k}{\|x_k\|} \in C$ and as C is closed it follows that $\lim_{k \to \infty} z_k = \bar{y} \in C$, so that $v^T\bar{y} < 0$, a contradiction.

THEOREM 5.3.7 A necessary condition for arbitrary-interval null-controllability of (A, B, Ω) is that the cone $C \triangleq \mathrm{Cl}(\mathrm{Conichull}(\mathrm{Convhull}(B\Omega)))$ is not pointed.

PROOF If the cone C is pointed there exists a $v \in R^n$ such that the hypothesis of Lemma 5.3.1 holds, which implies that the set Δ is compact. Clearly Δ is also convex and non-empty and is such that $0 \notin \Delta$. Theorem 5.3.3 then implies that the system (A, I, Δ) is not arbitrary-interval null-controllable. Noting that $\mathrm{Conichull}(\Delta) = C$ we see that Theorem 5.3.5 implies that the system (A, I, C) is not arbitrary-interval null-controllable. A further application of Theorem 5.3.5 yields the fact that (A, B, Ω) is not arbitrary-interval null-controllable.

The above theorem has the following implications.

COROLLARY 5.3.1 If $u \in \Omega \subseteq R^1$ and $0 \notin \mathrm{Int}(\mathrm{Convhull}(\Omega))$ then the system (A, B, Ω) is not arbitrary-interval null-controllable.

Note, however, that such a system can be null-controllable, cf. Section 5.2.1.

COROLLARY 5.3.2 If B has full rank then the cone C is not pointed if and only if the cone $\tilde{C} \triangleq \mathrm{Cl}(\mathrm{Conichull}(\mathrm{Convhull}(\Omega)))$ is not pointed.

Note also that regardless of the rank of B we always have that

Conichull(Convhull(BΩ)) = Conichull(B(Convhull(Ω))) = BConichull(Convhull(Ω)). If B has full rank then BCl(Conichull(Convhull(Ω))) = Cl(Conichull(Convhull(BΩ))).

Finally, Theorem 5.3.7 and Corollary 5.3.2 imply the following result.

COROLLARY 5.3.3 If B has full rank then a necessary condition for arbitrary-interval null-controllability of the system (A, B, Ω) is that the cone $\tilde{C}$ = Cl(Conichull(Convhull(Ω))) is not pointed.

5.3.4 Necessary and Sufficient Conditions - Finitely Generated Cones

When $\tilde{C}$ = Cl(Conichull(Convhull(Ω))) is a finitely generated cone we can deduce a necessary and sufficient condition for arbitrary-interval null-controllability of the system (A, B, Ω). We shall present an example which illustrates the use of the theorem, and another which illustrates that in general $\tilde{C}$ cannot be replaced by Ω in the theorem. A sufficient condition in the spirit of Theorem 5.1.3 is also presented.

THEOREM 5.3.8 Suppose that $\tilde{C}$ = Cl(Conichull(Convhull(Ω))) is a finitely generated cone. Then the system (A, B, Ω) is arbitrary-interval null-controllable if and only if there exists for each non-zero vector $v \in R^n$ (in fact, for each $v \in (B\tilde{C})'$, where $(\cdot)'$ is the polar of $(\cdot)$) an integer i(v), $0 \leqslant i(v) \leqslant n-1$ and a $u \in \Delta(v,i(v))$ such that $v^T A^{i(v)} Bu > 0$, where

$$\Delta(v,i) \triangleq \{u \in \tilde{C} | v^T A^j Bu=0,\ j=0,\dots,i-1\} \qquad (5.3.26)$$

$$\Delta(v,0) \triangleq \tilde{C}. \tag{5.3.27}$$

PROOF Sufficiency follows easily upon noting that

$$v^T e^{A\tau} Bu = v^T A^{i(v)} Bu \frac{\tau^{i(v)}}{i(v)!} + \text{higher-order terms in } \tau \tag{5.3.28}$$

for all $u \in \Delta(v,i(v))$. Indeed this equation and satisfaction of the conditions of the theorem imply that for each v there is a $t(v) > 0$ sufficiently small and a $u \in \tilde{C}$ such that $v^T e^{A\tau} Bu > 0$ for all τ, $0 < \tau \leqslant t(v)$. Theorems 5.3.4 and 5.3.5 then imply that the system (A, B, Ω) is arbitrary-interval null-controllable.

Suppose now that the system (A, B, Ω) is arbitrary-interval null-controllable. Then for a given $v \in R^n$ either there is a $u \in \tilde{C}$ such that $v^T Bu > 0$, in which case the conditions of the theorem are satisfied with $i(v) = 0$, or $v^T Bu \leqslant 0$ for all $u \in \tilde{C}$. If in addition to this last-mentioned case $\Delta(v,1) = \{0\}$ or $v^T ABu < 0$ for all $u \in \Delta(v,1)$ we have by the continuity in τ of $v^T e^{A\tau} Bu$ that there exists for each generating vector u_k of $\tilde{C}$, $k=1,\ldots,p$, a time $t_k(v)$ such that $v^T e^{A\tau} Bu_k \leqslant 0$ for all τ, $0 \leqslant \tau \leqslant t_k(v)$. Then, since each vector $u \in \tilde{C}$ is a non-negative combination of the generating vectors u_k, $k=1,\ldots,p$ we have $v^T e^{A\tau} Bu \leqslant 0$ for all $u \in \tilde{C}$ and $0 \leqslant \tau \leqslant t(v)$, where $t(v) = \min_k t_k(v)$. Hence Theorems 5.3.4 and 5.3.5 imply that the system is not arbitrary-interval null-controllable, a contradiction. This then implies that either $v^T ABu > 0$ for some $u \in \Delta(v,1)$ in which case $i(v) = 1$, or $i(v) > 1$. A similar argument may be used to investigate the larger possible values of $i(v)$. Clearly $i(v) \leqslant n-1$ as otherwise the Cayley-Hamilton theorem would imply that $v^T A^k Bu = 0$ for all $u \in \Delta(v,k)$ and all k and $\Delta(v,k) = \Delta(v,n)$ for all $k > n$. This in turn would imply, using a similar argument to that developed above which is based upon the finite generation of the cone $\tilde{C}$, that there is a time $t(v) > 0$ such that $v^T e^{A\tau} Bu \leqslant 0$ for all τ, $0 \leqslant \tau \leqslant t(v)$ and all $u \in \tilde{C}$, contradicting the assumed arbitrary-interval null-controllability of (A, B, Ω).

The following remarks which relate to Theorem 5.3.8 are important.

(i) When checking the conditions of Theorem 5.3.8 on a particular system (A, B, Ω) we need only consider those vectors $v \in R^n$ such that $v \in (\text{BConvhull}(\Omega))'$ where $(\cdot)'$ denotes the polar of $(\cdot)$.

(ii) If the cone $C = \text{Cl}(\text{Conichull}(\text{Convhull}(B\Omega)))$ is finitely generated, then Theorem 5.3.8 is applicable because the arbitrary-interval null-controllability of system (A, I, C) is equivalent to that of (A, B, $\tilde{C}$) and (A, B, Ω) - see Theorem 5.3.5.

(iii) If m = 2 or n = 2, Theorem 5.3.8 is always applicable because every cone in R^2 is finitely generated.

(iv) If the control constraint set Ω is a polyhedron, $\tilde{C}$ is finitely generated and Theorem 5.3.8 can be used.

The following theorem provides us with a set of sufficient conditions for arbitrary-interval null-controllability.

THEOREM 5.3.9 Suppose there exist $\Omega_s \subseteq \Omega$, $P \in R^{m' \times m}$ and $\tilde{B} \in R^{n \times m'}$ such that

$$Bu_s = \tilde{B}Pu_s \text{ for all } u_s \varepsilon \Omega_s \; ; \tag{5.3.29}$$

$$\text{the set } \{Pu_s | u_s \varepsilon \Omega_s\} \text{ contains the origin } Pu_s = 0 \text{ in its interior;} \tag{5.3.30}$$

$$\text{Rank}[\tilde{B}, A\tilde{B}, \ldots, A^{n-1}\tilde{B}] = n. \tag{5.3.31}$$

Then, the system (A, B, Ω) is arbitrary-interval null-controllable.

PROOF Naming the vector Pu_s as z we see that the system

$$\dot{x} = Ax + \tilde{B}z \tag{5.3.32}$$

$$z \; \varepsilon \; \Omega_z \triangleq \{Pu_s | u_s \; \varepsilon \; \Omega_s\} \tag{5.3.33}$$

satisfies the conditions of Theorem 5.3.8; i.e. the rank condition on the pair (A, $\tilde{B}$) and the fact that Ω_z contains zero in its interior are easily seen to be sufficient.

We remark that a special case of the conditions of Theorem 5.3.9 has been used previously in the literature. Specifically in [7] the minimum-time optimal control problem is considered and the following assumptions are made,

$$\Omega \subseteq R^n \text{ is a polyhedron.} \tag{5.3.34}$$

The so-called Pontryagin condition of general position is assumed, namely if b is a vector co-linear with an edge of the polyhedron Ω then

$$[b, Ab, \ldots, A^{n-1}b] = n \tag{5.3.35}$$

$$0 \in \Omega \text{ and } 0 \notin \text{vertex}(\Omega). \qquad (5.3.36)$$

We now show that these assumptions imply the existence of an Ω_s, P and $\tilde{B}$ as specified in Theorem 5.3.9. Indeed, if $0 \in \text{Int}(\Omega)$ we can set $\Omega_s = \Omega$, P = I, B = I. If $0 \notin \text{Int}(\Omega)$ then by (5.3.36) it must belong to an edge of Ω. We can then define Ω_s to be that edge, $\tilde{B}$ to be b (a vector co-linear with that edge) and P to be $\frac{b^T}{b^T b}$.

5.3.5 Examples

Our first example illustrates that in Theorem 5.3.8 the set $\tilde{C}$ cannot be replaced by Ω. We consider the following system:

$$A = \begin{pmatrix} 0 & 0 \\ 1 & 0 \end{pmatrix}, \quad B = \begin{pmatrix} 1 & 0 \\ 0 & 1 \end{pmatrix} \qquad (5.3.37)$$

$$\Omega = \{u = \begin{pmatrix} u_1 \\ u_2 \end{pmatrix} \in R^2 | u_1^2 + (u_2 - 1)^2 \leqslant 0\}. \qquad (5.3.38)$$

First we note that

$$\tilde{C} = \text{Cl}(\text{Conichull}(\text{Convhull}(\Omega))) = \{u | u_1 \in R^1, u_2 \geqslant 0\}. \qquad (5.3.39)$$

Now

$$v^T B u = v_1 u_1 + v_2 u_2 \qquad (5.3.40)$$

so that for each $v \in R^2$ for which $v_1 \neq 0$ or $v_2 > 0$ there is a $u \in \tilde{C}$ so that $v^T B u > 0$. The only case that remains is $v_1 = 0$, $v_2 < 0$. Here we note that

$$v^T Bu = 0 \text{ for } u_2 = 0,\ u_1 \text{ arbitrary} \tag{5.3.41}$$

so that

$$\Delta(v,1) = \{u | u_1 \in R^1,\ u_2 = 0\} \tag{5.3.42}$$

and there clearly exists a $u \in \Delta(v,1)$ such that

$$v^T ABu = -u_1 > 0. \tag{5.3.43}$$

Therefore the conditions of Theorem 5.3.8 are satisfied and the system (A, B, Ω) is arbitrary-interval null-controllable. However, if we replace $\tilde{C}$ by the set Ω we see that when $v_1 = 0$, $v_2 < 0$ there is no $u \in \Omega$ which causes $v^T Bu > 0$ and $\Delta(v,1)$ is just the set $\{0\}$, so that the conditions of the theorem cannot be satisfied.

Our next example is one which satisfies the conditions of Theorem 5.3.8 but not those of Theorem 5.3.9. The system equations are

$$\left.\begin{aligned} \dot{x}_1 &= x_1 + u_1 - u_2 \\ \dot{x}_2 &= -x_1 - x_2 + x_3 + u_1 \\ \dot{x}_3 &= -x_1 + x_2 - x_3 + u_3 - u_4 \end{aligned}\right\} \tag{5.3.44}$$

and

$$\Omega = \{u | u_i \geq 0,\ i=1,\ldots,4\} \tag{5.3.45}$$

so that

$$A = \begin{pmatrix} 1 & 0 & 0 \\ -1 & -1 & 1 \\ -1 & 1 & -1 \end{pmatrix}, \quad B = \begin{pmatrix} 1 & -1 & 0 & 0 \\ 1 & 0 & 0 & 0 \\ 0 & 0 & 1 & -1 \end{pmatrix}. \qquad (5.3.46)$$

We have that

$$v^T Bu = v_1(u_1 - u_2) + v_2 u_1 + v_3(u_3 - u_4). \qquad (5.3.47)$$

Clearly $v^T Bu > 0$ for some $u \in \Omega$ if $v_3 \neq 0$ and if $v_1 \neq 0$ and $v_2 = 0$. If $v_2 \neq 0$ then $v^T Bu$ fails to be positive for some $u \in \Omega$ only if $v_1 = v_3 = 0$, or $v_1 = -v_2$ and $v_3 = 0$, or $v_1 + v_2$ and v_1 are of opposite sign and $v_3 = 0$. The set $\Delta(v,1)$ for these three cases is, respectively

$$\{u | u_1 = 0, u_2 \geqslant 0, u_3 \geqslant 0, u_4 \geqslant 0\} \qquad (5.3.48)$$

$$\{u | u_1 \geqslant 0, u_2 = 0, u_3 \geqslant 0, u_4 \geqslant 0\} \qquad (5.3.49)$$

$$\{u | u_1 = u_2 = 0, u_3 \geqslant 0, u_4 \geqslant 0\}. \qquad (5.3.50)$$

Furthermore,

$$v^T ABu = v_1(u_1 - u_2) + v_2(-2u_1 + u_2 + u_3 - u_4) + v_3(u_2 - u_3 + u_4) \qquad (5.3.51)$$

and as $v_2 \neq 0$ and $v_3 = 0$ in all the above cases, (5.3.51) can be made positive for a $u \in \{u | u_1 = u_2 = 0, u_3 \geqslant 0, u_4 \geqslant 0\}$. Thus the conditions of Theorem 5.3.8 are satisfied and the system is arbitrary-interval null-controllable.

Now we note that (5.3.29) and (5.3.30) imply that Bu_s must change sign as a function of $u_s \in \Omega_s$ where Ω_s is a subset of

Ω. In our example this can happen only if $u_1 = u_2 = 0$. But then (5.3.44) is not even null-controllable, let alone arbitrary-interval null-controllable, and the sufficiency conditions of Theorem 5.3.9 cannot be satisfied.

5.3.6 Minimum Time Function

One of the most important and intensively studied problems in optimal control theory is that of steering the state of a linear dynamic system to the origin in minimum time. Usually one places enough restrictions and assumptions on the dynamic system and the constraint set Ω to ensure that the minimum time function $T(x)$, i.e. the minimum time to reach the origin from state x, is continuous in an open neighbourhood of the origin. In this section we show that the condition of arbitrary-interval null-controllability of the autonomous linear dynamic system is necessary and sufficient for the continuity of the minimum time function.

We need the following theorem which we quote in slightly modified form from [1].

THEOREM 5.3.10 Suppose that Ω is compact. If there exists a measurable control function $u_1(\cdot)$ with $u_1(t) \varepsilon \Omega$ which steers the state $x_0 \varepsilon R^n$ to the origin of the state space in time t_1 then there exists a measurable control function $u(\cdot)$ with $u(t) \varepsilon \Omega$ which steers x_0 to the origin in minimum time T.

Incidentally, note that the converse of Theorem 5.3.10 is trivially true, viz. if x_0 can be steered to the origin in minimum time T, then it can be steered to the origin in finite time $t_1 \geqslant T$!

The above theorem implies that the minimum time function $T(x)$ is defined in an open neighbourhood of $x = 0$ when Ω is compact if and only if (5.1.1) is null-controllable. However, null-controllability is not sufficient to ensure that $T(x)$ is continuous in an open neighbourhood of $x = 0$.

THEOREM 5.3.11 Suppose that Ω is compact. A necessary and sufficient condition for the minimum time function $T(x)$ to be continuous in an open neighbourhood of the origin of the state space is that (5.1.1) subject to (5.1.2) is arbitrary-interval null-controllable.

PROOF Note that $T(0) = 0$ and suppose that $T(x)$ is continuous in an open neighbourhood of the origin. Then the definition of continuity implies that for each $t > 0$ there is an open neighbourhood of the origin, say $V(t)$, such that all states in $V(t)$ can be steered to the origin in time t, which is arbitrary-interval null-controllability.

Suppose now that (5.1.1) subject to (5.1.2) is arbitrary-interval null-controllable. Then by Theorem 5.3.10 and subsequent remarks the minimum time function $T(x)$ is defined in an open neighbourhood of the origin. Furthermore arbitrary-interval null-controllability implies that for each $t > 0$ there is an open neighbourhood of the origin, say $V(t)$, such that all states in $V(t)$ can be steered to the origin in time t. Owing to the continuous dependence of the solutions of the differential equation on the initial values we know that given a point x in the domain of definition of $T(\cdot)$ there is a neighbourhood of x, say $\tilde{V}$, such that all points $y \in \tilde{V}$ can be steered in time $T(x)$ to $V(t)$. This then implies that all points $y \in \tilde{V}$ can be steered in time $T(x) + t$ to the origin.

It follows that for each $t > 0$ and x in the domain of $T(\cdot)$ there is a neighbourhood $\tilde{V}$ of x such that $T(y) - T(x) \leqslant t$ for all $y \varepsilon \tilde{V}$ which implies that $T(x)$ is upper-semicontinuous in an open neighbourhood of the origin. Now the set of states $\tilde{R}(t)$ which can be steered to the origin in time $t > 0$ is compact, convex, and varies continuously with t on $t \geqslant 0$, and $\tilde{R}(t_1) \subseteq \tilde{R}(t_2)$, $t_1 \leqslant t_2$ (cf. Theorem 5.1.4, the proof of Theorem 5.1.6 and Theorem 5.3.1). Moreover, if the minimum time to steer x to the origin is $T(x)$ we have that $x \varepsilon \partial\tilde{R}(T(x))$. Also such an x cannot belong to the sets $\tilde{R}(T(x)-t)$, $t > 0$ as $T(x)$ is the minimum time required to steer x to the origin. Therefore there exists for each $t > 0$ a neighbourhood $\tilde{V}$ of x such that $T(y) \geqslant T(x)-t$ for all $y \varepsilon \tilde{V}$, or $T(x) - T(y) \leqslant t$ for all $y \varepsilon \tilde{V}$, which implies that $T(x)$ is lower-semicontinuous in an open neighbourhood of the origin. The fact that $T(x)$ is both upper- and lower-semicontinuous in an open neighbourhood of the origin implies that it is continuous in an open neighbourhood of the origin.

The above theorem implies that Pontryagin's conditions (5.3.34) - (5.3.36) which are sufficient for arbitrary-interval null-controllability also guarantee the continuity of the minimum time function $T(x)$. It also follows that our necessary and sufficient conditions for arbitrary-interval null-controllability are the <u>minimal</u> conditions required for the continuity of the minimum time function.

5.3.7 Further Necessary Conditions and Sufficient Conditions

In this section we present necessary conditions and sufficient conditions due to M. Pachter for arbitrary-interval null-controllability which do not require the assumption that Cl(Conichull(Convhull(Ω))) is finitely generated. Furthermore they are of geometric character and yield insight into the geometric aspects of arbitrary-interval null-controllability.

We require the following preliminary lemma and subsequent observations.

LEMMA 5.3.2 Suppose that C is a convex cone in R^n and that S is a subspace contained in C. Then

$$C = S + S^{\perp} \cap C. \tag{5.3.52}$$

PROOF Clearly $S + S^{\perp} \cap C \subseteq (S+S^{\perp}) \cap C = C$. To prove the converse we suppose that $c \in C$ and $c = c_1 + c_2$ where $c_1 \in S$ and $c_2 \in S^{\perp}$. Then $c_2 = c - c_1$, and $-c_1 \in S$ since S is a subspace. As $S \subseteq C$ we have $-c_1 \in C$ and the convexity of C implies that $c_2 \in C$. Consequently we have that $c \in S + S^{\perp} \cap C$ which in turn implies that $C \subseteq S + S^{\perp} \cap C$.

From this point on we suppose that S is the largest subspace contained in $C \triangleq$ Cl(Conichull(Convhull(BΩ))). Clearly we have that

$$S = C \cap (-C) \tag{5.3.53}$$

and an algorithm for computing S is available [8]. Note that Theorem 5.3.7 implies that {0} is a strict subset of S. Lemma 5.3.2 implies that C can be written as

$$C = S + S^{\perp} \cap C \tag{5.3.54}$$

where $S^{\perp} \cap C$ is a closed, convex, pointed cone. That $S^{\perp} \cap C$ is pointed follows from the fact that $(S^{\perp} \cap C) \cap [-(S^{\perp} \cap C)] = S^{\perp} \cap C \cap (-S^{\perp}) \cap (-C) = S^{\perp} \cap C \cap (-C) \cap (-S^{\perp}) = S^{\perp} \cap S \cap S^{\perp} = \{0\}$. Furthermore we note that $(B\mathrm{Convhull}(\Omega))' \subseteq S^{\perp}$ and $(B\mathrm{Convhull}(\Omega))' = (\mathrm{Cl}(\mathrm{Conichull}(\mathrm{Convhull}(B\Omega))))'$.

Before presenting our main theorem we define the following notation. Given a set $X \subseteq S^{\perp}$ we denote by X^{o} the interior of the set X relative to the subspace $S^{\perp}$, plus the set $\{0\}$. We denote by $\{A|S\}$ the smallest subspace invariant under A which contains the subspace S. It is well known that $\{A|S\} = S + AS + \dots + A^{n-1}S$.

THEOREM 5.3.12 A necessary condition for arbitrary-interval null-controllability of (5.1.1) subject to (5.1.2) is that

$$((B\mathrm{Convhull}(\Omega))')^{o} \cap \{A|S\}^{\perp} = \{0\}. \tag{5.3.55}$$

A sufficient condition for arbitrary-interval null-controllability is that

$$(B\mathrm{Convhull}(\Omega))' \cap \{A|S\}^{\perp} = \{0\} \tag{5.3.56}$$

which is equivalent, as $0 \in B\mathrm{Convhull}(\Omega)$ - cf. Theorem 5.3.4, to the sufficient condition

$$0 \in \mathrm{Int}(\mathrm{Convhull}(B\Omega)+\{A|S\}). \tag{5.3.57}$$

PROOF We first consider necessity. Let $v \in ((B\mathrm{Convhull}(\Omega))')^{o}$, with $\|v\| = 1$. It follows that $v^{T}c < 0$ for all $c \in C \cap S^{\perp}$, $c \neq 0$ so that Lemma 5.3.1 implies that the set

$\Delta_v \overset{\Delta}{=} \{y|v^T y = -1, y \in C \cap S^\perp\}$ is compact and convex and it is clear that Conichull$(\Delta_v) = C \cap S^\perp$. Then it follows because of the compactness of Δ_v that there is a $t > 0$ such that $v^T e^{A\tau} y \leqslant 0$ for all τ, $0 \leqslant \tau \leqslant t$ and for all $y \in \Delta_v$. Consequently by Theorem 5.3.2 the system (A, I, Δ_v) is not arbitrary-interval null-controllable and by Theorem 5.3.5 this is true also of the system $(A, I, C \cap S^\perp)$, which implies that there is a $t > 0$ such that $v^T e^{A\tau} y \leqslant 0$ for all τ, $0 \leqslant \tau \leqslant t$ and for all $y \in C \cap S^\perp$. Now assume that v belongs to a subspace of $S^\perp$ invariant under A^T. Then $e^{A^T\tau} v \in S^\perp$ for all $\tau \geqslant 0$ so that $v^T e^{A\tau} s = 0$ for all $s \in S$ and all $\tau \geqslant 0$. Combining the two results with the aid of Lemma 5.3.2 yields $v^T e^{A\tau} c \leqslant 0$ for all τ, $0 \leqslant \tau \leqslant t$ and all $c \in C$ so that Theorem 5.3.2 implies that the system is not arbitrary-interval null-controllable. Thus it is necessary for arbitrary-interval null-controllability that $((B\text{Convhull}(\Omega))')^o \cap$ (largest subspace invariant under A^T contained in $S^\perp$) $= \{0\}$. Reference [9] provides the last step, viz. the largest subspace invariant under A^T contained in $S^\perp$ is $\{A|S\}^\perp$.

Now we go on to sufficiency. We assume that $0 \in \text{Int}(\text{Convhull}(B\Omega) + \{A|S\})$ which is equivalent to $(\text{Convhull}(B\Omega) + \{A|S\})' = \{0\}$. A set-theoretic identity then implies that

$$(\text{Convhull}(B\Omega))' \cap \{A|S\}^\perp = \{0\}. \qquad (5.3.58)$$

Next we note that if (5.3.58) holds then $S \neq \{0\}$. Indeed, if $S = \{0\}$ we would have from (5.3.58) that $(\text{Convhull}(B\Omega))' = \{0\}$

which is equivalent to $0 \in \text{Int}(\text{Convhull}(B\Omega))$. This then implies that $0 \in \text{Int}(\text{Convhull}(\Omega))$ and $R(B) = R^n$ so that $S = R^n$, which is a contradiction, which means that $\{0\}$ is a strict subset of S. As mentioned in the proof of necessity, $\{A|S\}^{\perp}$ is the largest subspace invariant under A^T contained in $S^{\perp}$, which implies that for all $v \in (\text{Convhull}(B\Omega))'$ there is an i, $0 < i \leqslant n-1$ and an $\bar{s} \in S \subseteq C$ such that $v^T A^j s = 0$ for all $s \in S$ and $j=0,1,\ldots,i-1$ and $v^T A^i \bar{s} > 0$. Also, if $v \notin (\text{Convhull}(B\Omega))'$ then there is a $y \in C$ such that $v^T y > 0$. In either case, then, we have that for each v there exists a $y \in C$ and a $t > 0$ such that $v^T e^{A\tau} y > 0$, $0 \leqslant \tau \leqslant t$ which implies that the system (A, I, C) is arbitrary-interval null-controllable. Theorem 5.3.5 then implies that the system (A, I, $B\Omega$) is arbitrary-interval null-controllable, which implies that the system (A, B, Ω) is arbitrary-interval null-controllable. Conditions (5.3.56) and (5.3.57) are equivalent if $0 \in \text{Convhull}(B\Omega)$ because in that case $(\text{Convhull}(B\Omega) + \{A|S\})' = (\text{Convhull}(B\Omega))' \cap \{A|S\}^{\perp}$ - see Section 5.A.

The following corollaries follow from the above theorem.

COROLLARY 5.3.4 A necessary condition for arbitrary-interval null-controllability of (5.1.1) subject to (5.1.2) is that Rank(Q) = n.

COROLLARY 5.3.5 Assume that $0 \in \text{Int}(\text{Convhull}(\Omega))$. Then a necessary and sufficient condition for arbitrary-interval null-controllability of (5.1.1) subject to (5.1.2) is that Rank(Q) = n.

Furthermore we note that the sufficient conditions of Theorem 5.3.12 are less stringent than those of Theorem 5.3.9. Indeed, returning to the example of Section 5.3.5 we see that

$$S = \text{linear span}\left(\begin{pmatrix} 0 \\ 0 \\ 1 \end{pmatrix}\right) \tag{5.3.59}$$

$$\text{BConvhull}(\Omega) = \{y \in R^3 | y_1 \leqslant y_2,\ y_2 \geqslant 0,\ y_3 \in R^1\} \tag{5.3.60}$$

and

$$\{A|S\} = \text{linear span}\left(\begin{pmatrix} 0 \\ 0 \\ 1 \end{pmatrix}, \begin{pmatrix} 0 \\ 1 \\ -1 \end{pmatrix}\right). \tag{5.3.61}$$

It follows that

$$\{A|S\}^{\perp} = \text{linear span}\left(\begin{pmatrix} 1 \\ 0 \\ 0 \end{pmatrix}\right) \tag{5.3.62}$$

and $\begin{pmatrix} 1 \\ 0 \\ 0 \end{pmatrix}$ does not belong to (BConvhull(Ω))'. Therefore condition (5.3.56) is satisfied and the system (5.3.44) subject to (5.3.45) is arbitrary-interval null-controllable.

We note, however, that there is a gap between the necessary condition and the sufficient conditions of Theorem 5.3.12. Indeed, if we set

$$A = \begin{pmatrix} 0 & 0 & 0 & 1 \\ 0 & 1 & 0 & 0 \\ 0 & 1 & 0 & 1 \\ 0 & 0 & 0 & 0 \end{pmatrix}, \quad B = I_4 \tag{5.3.63}$$

and

$$\Omega = \text{linear span}\left(\begin{pmatrix} 0 \\ 0 \\ 0 \\ 1 \end{pmatrix}\right) + \{y \in R^4 | y_1 \geqslant 0,\ y_2^2 + y_3^2 \leqslant y_1^2,\ y_4 = 0\} \tag{5.3.64}$$

we find that the sufficient conditions of Theorem 5.3.12 are not satisfied but it can be verified direct using Theorem 5.3.4 that the system (A, B, Ω) is arbitrary-interval null-controllable.

5.3.8 Further Special Cases

THEOREM 5.3.13 Assume that $S^{\perp} \subseteq R^2$. Then a necessary and sufficient condition for arbitrary-interval null-controllability of (5.1.1) subject to (5.1.2) is that

$$(\text{Convhull}(B\Omega))' \cap \{A|S\}^{\perp} = \{0\} \tag{5.3.65}$$

which is equivalent, because of our assumption that $Bu = 0$ for some $u \in \Omega$, to

$$0 \in \text{Int}(\text{Convhull}(B\Omega)+\{A|S\}). \tag{5.3.66}$$

PROOF We need only prove the necessity of the condition, as sufficiency is already proved in Theorem 5.3.12. In fact all we need to prove is that all vectors in the boundary of $(\text{Convhull}(B\Omega))'$ are not in $\{A|S\}^{\perp}$. Thus, suppose $v \in \partial(\text{Convhull}(B\Omega))'$. If $v \in \{A|S\}^{\perp}$ then since $S^{\perp} \subseteq R^2$ either v is an eigenvector of A^T or $S^{\perp}$ is invariant under A^T. In the former case $v^T e^{At} c = e^{\lambda t} v^T c \leqslant 0$ for all $c \in C$ and $t \geqslant 0$ so that we would not have arbitrary-interval null-controllability. Assume then that $S^{\perp}$ is invariant under A^T. Then those vectors $v \in (\text{Convhull}(B\Omega))'$ such that $v^T c < 0$ for all $c \in C \cap S^{\perp}$, $c \neq 0$, are also in $\{A|S\}^{\perp}$ so that, following the proof of necessity in Theorem 5.3.12, there exists $t > 0$ such that $v^T e^{A\tau} c \leqslant 0$ for all $c \in C$ and τ, $0 \leqslant \tau \leqslant t$ so that we would not have arbitrary-interval null-controllability.

COROLLARY 5.3.6 Suppose that n = 3, i.e. that $x \in R^3$. Then (5.3.65) and (5.3.66) are necessary and sufficient for arbitrary-interval null-controllability of (5.1.1) subject to (5.1.2).

PROOF Again we need only prove necessity. Arbitrary-interval null-controllability implies that C is not pointed so that the dimension of the largest subspace S is greater than or equal to 1. Hence the dimension of $S^{\perp}$ is less than or equal to n - 1 = 2 and the corollary follows from Theorem 5.3.13.

We remark that Corollary 5.3.6 allows us to always use Theorem 5.3.8 when $n \leqslant 3$. Indeed, as stated above, the dimension of S is not less than one and that of $S^{\perp}$ is not greater than 2. As every cone in R^2 is finitely generated so is $S^{\perp} \cap C$ and as S is a subspace the cone C is finitely generated.

5.4 Conclusion

In the introductory section of this chapter we reviewed well-known results on controllability and null-controllability of autonomous linear systems and described and discussed the most important properties of reachable sets. Section 2 was devoted to a study of null-controllability when Ω and its convex hull do not contain zero in their interior. Specifically we proved Brammer's important theorem via Hãjek's approach.

In Section 3 we introduced the important notion of arbitrary-interval null-controllability and proved a number of preliminary necessary and sufficient conditions. Then we proved that in questions of arbitrary-interval null-controllability Ω can be replaced by the cone $\tilde{C} = Cl(Conichull(Convhull(\Omega)))$ and this led to the development of necessary conditions, such as C should be not pointed, and necessary and sufficient conditions when $\tilde{C}$ is a finitely generated cone. We showed by means of a sufficient condition that Pontryagin's condition of general position implies arbitrary-interval null-controllability. We also presented a number of illustrative examples. Next we proved the important result that the minimum time function is continuous in an open neighbourhood of the origin of the state space if and only if the system (A, B, Ω), Ω compact, is arbitrary-interval null-controllable. This justifies further the introduction of the notion of arbitrary-interval null-controllability. Following this we presented necessary conditions and sufficient conditions of a geometric nature which are due to M. Pachter. These conditions do not require that the cone C be finitely generated.

The chapter provides a rather complete treatment of the 'controllability' of constrained, linear, autonomous continuous-time, dynamic systems.

5.5 References

[1] LEE, E.B. & MARKUS, L. Foundations of Optimal Control Theory. Wiley, New York, 1967.

[2] BRAMMER, R.F. Controllability in Linear Autonomous Systems with Positive Controllers. SIAM J. Control, 10, 1972, pp. 339-353.

[3] SAPERSTONE, S.H. & YORKE, J.A. Controllability of Linear Oscillatory Systems Using Positive Controls. SIAM J. Control, 9, 1971, pp. 253-262.

[4] HÀJEK, O. A Short Proof of Brammer's Theorem. Unpublished preprint, 1975.

[5] HEYMANN, M. & STERN, R.J. Controllability of Linear Systems with Positive Controls: Geometric Considerations. J. Math. Anal. Appl., to appear.

[6] PACHTER, M. & JACOBSON, D.H. Conditions for the Controllability of Constrained Linear Autonomous Systems on Time Intervals of Arbitrary Length. CSIR Special Report WISK 210, July 1976, 23 p.

[7] BOLTYANSKII, V.E. Mathematical Methods of Optimal Control. Holt, Rinehart, Winston, New York, 1971, pp. 119-120.

[8] ECKHARDT, V. Theorems on the Dimension of Convex Sets. Linear Algebra and its Applications, 12, 1975, pp. 63-76.

[9] WONHAM, W.M. On the Matrix Riccati Equation of Stochastic Control. SIAM J. Control, 6, 1968, pp. 681-697.

[10] ROCKAFELLAR, R.T. Convex Analysis. Princeton University Press, 1970.

5.A Appendix

The following definitions and facts are used in Chapter 5; for further details see [10].

The set $C \subseteq R^n$ is a cone if for all $c \in C$ and all scalars $\alpha \geqslant 0$ we have $\alpha c \in C$.

A cone C is convex if and only if $C + C \subseteq C$.

A cone C is pointed if $C \cap (-C) = \{0\}$.

A cone C is pointed if and only if there is a $v \in R^n$ such that $v^T c < 0$ for all $c \in C$, $c \neq 0$.

The convex hull of a set $\Omega \subseteq R^m$, denoted Convhull(Ω), is the smallest convex set which contains Ω.

The conic hull of a set $\Omega \subseteq R^m$, denoted Conichull(Ω), is the smallest cone with vertex at the origin which contains Ω.

The closure of a set $\Omega \subseteq R^m$, denoted Cl(Ω), is the smallest closed set containing Ω.

If the set Ω is bounded, so is Convhull(Ω); if the set Ω is closed, so is Convhull(Ω). Thus Ω compact implies that Convhull(Ω) is compact. If the set Ω is convex, so is the set Conichull(Ω).

Let B be a map $B:R^m \to R^n$. We denote by $B(\Omega)$ or $B\Omega$ the image under B of the set Ω.

Let $T \subseteq R^n$. Then T', the polar of T, is defined as $T' = \{x \in R^n | x^T y \leq 0 \text{ for all } y \in T\}$. The set T' is a closed and convex cone, and if T is a subspace then $T' = T^{\perp}$. We note that $T \subseteq T''$ where $T'' \triangleq (T')'$. If T is convex then $T' = \{0\}$ if and only if $0 \in \text{Int}(T)$.

Let $T, V \subseteq R^n$. Then $T' \cap V' \subseteq (T+V)'$ and if $0 \in T \cap V$ then $T' \cap V' = (T+V)'$.

6. NEW APPROACHES TO FUNCTION MINIMIZATION

6.1 Introduction

The problem of function minimization is to find the minimizer $\hat{x} \in R^n$ of a scalar function

$$f(x), \quad f:R^n \to R^1. \tag{6.1.1}$$

Since the emergence of Fletcher and Powell's algorithm [1] in 1963 quasi-Newton methods have dominated the field of numerical algorithms for minimizing differentiable functions. Quasi-Newton methods iteratively approach the minimizer (or at least a stationary point) of f(x) via a recurrence relation of the type

$$x_{i+1} = x_i - \rho_i S_i g_i \tag{6.1.2}$$

where $\rho_i \in [0,\infty)$ is the scalar 'step length' at iteration i, S_i is a symmetric n x n matrix which approximates the inverse of the second-derivative matrix $f_{xx}(x)$ of f(x) at the point x_i, and $g_i \in R^n$ is the transpose of the gradient row-vector $f_x(x)$ evaluated at the point x_i. The name 'quasi-Newton' arises from the similarity of (6.1.2) to Newton's iteration formula

$$x_{i+1} = x_i - [f_{xx}(x_i)]^{-1} g(x_i) \tag{6.1.3}$$

which in turn follows from minimizing the quadratic terms in the Taylor series expansion of f(x) in the neighbourhood of x_i, viz.

$$f(x) = f(x_i) + f_x(x_i)(x-x_i) + \tfrac{1}{2}(x-x_i)^T f_{xx}(x_i)(x-x_i)$$

$$+ \text{ higher-order terms.} \qquad (6.1.4)$$

Basically, then, most algorithms for function minimization are derived from a quadratic approximation of the objective function, $f(x)$.

In this chapter we present two new approaches to function minimization which depart from quadratic approximations, or models, and which therefore exhibit characteristics and advantages not shared by conventional algorithms. We shall assume throughout that $f(x)$ is twice-continuously differentiable in R^n, though in many instances this assumption may be weakened to once-continuous differentiability.

6.2 Homogeneous Models

In 1971 a simple observation led to a useful generalization of the quadratic model [1]. Specifically we note that the quadratic function,

$$f(x) = \tfrac{1}{2}(x-\hat{x})^T Q(x-\hat{x}) + \bar{\omega} \qquad (6.2.1)$$

where Q is an $n \times n$ symmetric positive semi-definite matrix and $\bar{\omega}$ is a real scalar, may be written as

$$f(x) = \tfrac{1}{2} f_x(x)(x-\hat{x}) + \bar{\omega} \qquad (6.2.2)$$

so that multiplying through by the factor 2 we obtain

$$2f(x) = f_x(x)(x-\hat{x}) + 2\bar{\omega} \ . \qquad (6.2.3)$$

We now generalize this model via the following definition.

DEFINITION 6.2.1 A function f(x), $f:R^n \to R^1$ is said to be homogeneous of degree γ if it satisfies the equation

$$\gamma f(x) = f_x(x)(x-\hat{x}) + \omega \qquad (6.2.4)$$

where $\gamma \in R^1$.

Clearly the quadratic function (6.2.1) is homogeneous of degree $\gamma = 2$, but there are many other functions which are homogeneous. Notably the function

$$f(x) = [(x-\hat{x})^T Q(x-\hat{x})]^p, \quad p > 0 \qquad (6.2.5)$$

is homogeneous of degree $\gamma = 2p$. Quasi-Newton methods behave particularly poorly when applied to (6.2.5) because the second derivative matrix at $\hat{x}$, $f_{xx}(\hat{x})$, is identically zero whenever $p > 1$.

6.2.1 Relation to Newton's Formula

It is instructive to explore the relationship between Newton's formula and the homogeneous function. If we differentiate (6.2.4) with respect to x we obtain

$$g(x) = \frac{1}{\gamma} f_{xx}(x)(x-\hat{x}) + \frac{1}{\gamma} g(x) \qquad (6.2.6)$$

where $g(x) = f_x^T(x)$, which yields the following expression for $\hat{x}$, the assumed minimizer of f(x),

$$\hat{x} = x - (\gamma-1)[f_{xx}(x)]^{-1} g(x). \qquad (6.2.7)$$

Thus assuming that $f_{xx}(x)$ is invertible we can locate the minimizer of a homogeneous function by scaling the 'Newton step'

$$-\ [f_{xx}(x)]^{-1}g(x) \qquad (6.2.8)$$

by the factor $(\gamma-1)$. Incidentally a similar result was obtained in 1870 by Schröder for the closely related problem of finding multiple roots of a polynomial [2].

The above relationship between the scaled 'Newton step' and the homogeneous function implies, if the second derivative matrix is available and is positive definite for all $x \in R^n$, that the iteration formula

$$x_{i+1} = x_i - \rho_i [f_{xx}(x_i)]^{-1}g(x_i) \qquad (6.2.9)$$

with ρ_i chosen so as to minimize $f(x_i-\rho_i [f_{xx}(x_i)]^{-1}g(x_i))$ would yield an effective algorithm which would locate the minimum of a homogeneous function in one step. Indeed this has been borne out in practice. On the other hand this feature is lost if $f_{xx}(x)$ is not explicitly available - except on quadratic functions when quasi-Newton methods accurately estimate the then constant second derivative matrix in n steps.

6.2.2 Algorithms based upon Homogeneous Functions

The potential usefulness of the homogeneous function (6.2.4) lies in the fact that the unknown parameters γ, $\hat{x}$ and ω appear linearly. If we define

$$v \triangleq x^T g(x) \qquad (6.2.10)$$

$$y^T \triangleq [g^T(x), f(x), -1] \qquad (6.2.11)$$

$$\alpha^T \triangleq [\hat{x}^T, \gamma, \omega] \qquad (6.2.12)$$

we can write (6.2.4) as

$$y^T \alpha = v \qquad (6.2.13)$$

so that (n+2) linearly independent 'observations' y and the corresponding values of v determine the vector of unknowns α. Thus the location of the minimum of a homogeneous function, its degree of homogeneity and its scaled minimum value can be determined from (n+2) data which require only the evaluation of f(x) and g(x). When seeking the minimizer of a non-homogeneous function we specify an updating formula which generates successive approximations to $\hat{x}$, the minimizer of f(x). Details of two algorithms which accomplish this via Householder's updating formula are in [1], [3]. Good numerical results were obtained when using each algorithm.

Recently Kowalik and Ramakrishnan [4] have taken a close look at the homogeneous approach and have made the important observation that the updating methods used in [1], [3] can be substantially improved upon. In fact they replace the Householder updating formula by a semi-triangular factorization which is numerically stable (in contrast to the Householder formula) provided that a pivoting strategy is used in the process of updating the factors. Their numerically stable realization performs substantially better than the original homogeneous algorithms [1], [3] and suggests that the homogeneous model will find widespread application.

Two important other advantages of the model (6.2.4) are that it is not necessary to choose the 'step-size' so as to minimize $f(x)$ along the search direction, and that the second derivative matrix $f_{xx}(x)$ need not be positive definite.

6.3 Differential Descent

The iteration formula (6.1.2) which is used in almost all quasi-Newton algorithms is rather simple. Specifically one determines a suitable step size ρ_i along the fixed direction $S_i g_i$.

In this section we introduce a new approach which generalizes this iteration formula to the form

$$x_{i+1} = x_i + p_i(x_i, \hat{t}) \tag{6.3.1}$$

where $\hat{t} \in R^1$ plays the role of a 'generalized' step length. The search 'direction' $p_i(x_i, t)$ is a non-linear function of the step length t.

6.3.1 Outline of the Approach

We consider the vector differential equation [5]

$$\frac{dx}{dt} = - g(x) \tag{6.3.2}$$

which is non-linear whenever $f(x)$ is non-quadratic. The solution of (6.3.2) for different initial conditions defines a family of curves $x(\cdot)$ which are normal to the contours of constant function value, viz. $f(x) = c$.

If $\hat{x}$ is the unique minimizer of $f(x)$ and if $|f(x)| \to \infty$ as $\|x\| \to \infty$ then the function

$$V(x) \triangleq f(x) - f(\hat{x}) \tag{6.3.3}$$

is positive definite and radially unbounded. Furthermore its derivative with respect to time is given by

$$\frac{dV}{dt}(x) = -\, g^T(x)g(x) \tag{6.3.4}$$

which is negative whenever $x \neq \hat{x}$. It then follows from a well-known stability theorem of Liapunov that $x(t) \to \hat{x}$ as $t \to \infty$. Consequently we wish to determine the asymptotic solution of the differential equation (6.3.2) having started at $t = 0$ from the initial condition

$$x(0) = x_i \tag{6.3.5}$$

where x_i is the best known approximation to $\hat{x}$.

Naturally we could integrate (6.3.2) from the initial condition (6.3.5) using any one of a number of integration techniques. However, for all interesting functions $f(x)$ it turns out that (6.3.2) is a stiff differential equation and consequently most integration techniques perform satisfactorily only if small steps in t are taken. Botsaris [6] provides details of suitable numerical integration methods which take large steps while we pursue a different approach here.

We approximate $g(x)$ in the neighbourhood of x_i by the

expression

$$g_i + H_i(x-x_i) \tag{6.3.6}$$

where

$$H_i \triangleq f_{xx}(x_i). \tag{6.3.7}$$

Equation (6.3.2) is therefore replaced by

$$\frac{dx}{dt} = - g_i - H_i x + H_i x_i, \quad x(0) = x_i \tag{6.3.8}$$

which has the solution

$$x(t) = x_i + (e^{-tH_i}-I)H_i^{-1}g_i. \tag{6.3.9}$$

Now if we find the minimum of $f(x(t))$ with respect to t (possibly at $\hat{t} = \infty$) we obtain the new estimate of the minimizer as

$$x_{i+1} = x_i + (e^{-\hat{t}H_i}-I)H_i^{-1}g_i. \tag{6.3.10}$$

Actually equation (6.3.10) is not used direct owing to the fact that it requires that H_i be invertible. We can, however, put it in a different form by noting that if $u_1,\ldots,u_n$ are the normalized eigenvectors and $\lambda_1,\ldots,\lambda_n$ the eigenvalues of the symmetric matrix H_i then

$$(e^{-tH_i}-I)H_i^{-1} = \sum_{j=1}^{n} \frac{e^{-\lambda_j t}-1}{\lambda_j} u_j u_j^T \tag{6.3.11}$$

so that (6.3.9) becomes

$$x(t) = x_i + \left(\sum_{j=1}^{n} \frac{e^{-\lambda_j t} - 1}{\lambda_j} u_j u_j^T \right) g_i. \qquad (6.3.12)$$

If $\lambda_j = 0$ we can by L'Hospital's rule replace the undefined term $\frac{e^{-\lambda_j t}-1}{\lambda_j}$ in equation (6.3.12) by $-t$.

The following lemma is then of interest and importance.

LEMMA 6.3.1 The matrix $\left(\sum_{j=1}^{n} \frac{e^{-\lambda_j t}-1}{\lambda_j} u_j u_j^T \right)$, with $\frac{e^{-\lambda_j t}-1}{\lambda_j}$ replaced by $-t$ if $\lambda_j = 0$, is negative definite for $t > 0$ regardless of the value of λ_j.

PROOF If $\lambda_j > 0$ then for $t > 0$, $e^{-\lambda_j t} < 1$ so that $\frac{e^{-\lambda_j t}-1}{\lambda_j} < 0$. On the other hand if $\lambda_j < 0$ then for $t > 0$, $e^{-\lambda_j t} > 1$ so that again $\frac{e^{-\lambda_j t}-1}{\lambda_j} < 0$. Finally, if $\lambda_j = 0$ we use L'Hospital's rule to yield

$$\lim_{\lambda_j \to 0} \frac{e^{-\lambda_j t}-1}{\lambda_j} = \lim_{\lambda_j \to 0} \frac{\frac{d}{d\lambda_j}(e^{-\lambda_j t}-1)}{\frac{d}{d\lambda_j}(\lambda_j)} = -t \qquad (6.3.13)$$

which is also negative for $t > 0$, and the lemma is proved.

LEMMA 6.3.2 There exists a $\delta > 0$ such that $f(x(t)) < f(x_i)$ for all $t \in (0,\delta)$. Moreover, for small t, (6.3.12) is approximately

$$x(t) \approx x_i - tg_i \tag{6.3.14}$$

and provided that H_i is positive definite we have that for large t

$$x(t) \approx x_i - H_i^{-1}g_i. \tag{6.3.15}$$

Therefore for small t, (6.3.12) yields a 'gradient step' and for large t, a 'Newton step'.

PROOF It is easy to verify that $\frac{df}{dt}$ evaluated at t = 0 is simply equal to $-g_i^T g_i$. Consequently we have that

$$\lim_{t \to 0} \frac{f(x(t))-f(x_i)}{t} + g_i^T g_i = 0. \tag{6.3.16}$$

The existence of $\frac{d}{dt}f(x(t))$ then implies that we can find a $\delta > 0$ so that for all $t \in (0,\delta)$

$$\frac{f(x(t))-f(x_i)}{t} + g_i^T g_i < g_i^T g_i \tag{6.3.17}$$

which, since $t > 0$, implies that

$$f(x(t)) - f(x_i) < 0 \text{ for all } t \in (0,\delta). \tag{6.3.18}$$

Now for small t we have that $\frac{e^{-\lambda_j t}-1}{\lambda_j} \approx -t$ and as the u_j are orthogonal and normalized, (6.3.14) results. If H_i is positive definite $e^{-H_i t} \to 0$ as $t \to \infty$ so that (6.3.15) follows from (6.3.9).

Lemmas 6.3.1. and 6.3.2 demonstrate the advantages of using the search curve (6.3.12) in place of the conventional quasi-Newton search vector

$$x(t) = x_i - tS_i g_i. \qquad (6.3.19)$$

The disadvantage is that it is necessary to find the normalized eigenvectors and the eigenvalues of H_i, the second derivative matrix of $f(x)$ evaluated at x_i [5]. In a recent article Botsaris [7] has overcome this disadvantage by developing formulae which recursively estimate the required eigenvectors and eigenvalues from the already collected data f_i, g_i, f_{i-1}, g_{i-1}, ... The resulting algorithm has proved to be very powerful indeed [8], outperforming Fletcher's algorithm [9].

6.4 Conclusion

In this brief chapter we indicated two new approaches to function minimization which depart from the conventional use of quadratic models and quasi-Newton formulae. Details of the algorithms which are based upon these approaches are not given, as these are not within the scope of the present volume, which is dedicated to conceptual 'quadratic extensions' rather than descriptions of implementable algorithms.

6.5 References

[1] JACOBSON, D.H. & OKSMAN, W. An Algorithm that Minimizes Homogeneous Functions of n Variables in n+2 Iterations and Rapidly Minimizes General Functions. J. Math. Anal. Appl., 38, 1972, pp. 535-552.

[2] OSTROWSKI, A.M. Solutions of Equations and Systems of Equations. Academic Press, New York, 1966, p. 67.

[3] JACOBSON, D.H. & PELS, L.M. A Modified Homogeneous Algorithm for Function Minimization. J. Math. Anal. Appl., 46, 1974, pp. 533-541.

[4] KOWALIK, J.S. & RAMAKRISHNAN, K.G. A Numerically Stable Optimization Method Based on a Homogeneous Function. Math. Prog., to appear.

[5] BOTSARIS, C.A. & JACOBSON, D.H. A Newton-Type Curvilinear Search Method for Optimization. J. Math. Anal. Appl., 54, 1976, pp. 217-229.

[6] BOTSARIS, C.A. A Class of Methods for Unconstrained Minimization based on Stable Numerical Integration Techniques. CSIR Special Report WISK 215, July, 1976.

[7] BOTSARIS, C.A. A Curvilinear Optimization Method based upon Iterative Estimation of the Eigensystem of the Hessian Matrix. CSIR Special Report WISK 217, July, 1976.

[8] BOTSARIS, C.A. A Comparison Study of a New Algorithm for Function Minimization. CSIR Special Report WISK 214, July, 1976.

[9] FLETCHER, R. A New Approach to Variable Metric Algorithms. Computer J., 13, 1970, pp. 317-322.

7. CONCLUSION

The purpose of this monograph is to demonstrate that non-trivial extensions can be made to 'linear-quadratic theory'.

In Chapter 2 we presented our results on exponential performance criteria and non-linear dynamic systems of special structure in both stochastic and deterministic settings.

The theory of non-convex quadratic forms presented in Chapter 3 was developed along lines which are not well known but which are direct and simple. This led to extensions of Finsler's theorem and related results and finally to sufficient conditions for the existence of finite escape times in solutions of systems of quadratic differential equations.

Chapter 4 was devoted to extensions of our sufficiency conditions for non-negativity of quadratic functionals to constrained and non-linear cases. In the case of linear inequality constraints on the control and state variables this approach led to a novel Riccati differential equation.

The study of arbitrary-interval null-controllability presented in Chapter 5 led to necessary and sufficient conditions for the continuity of the minimum time function. Our necessary and sufficient conditions for arbitrary-interval null-controllability are mainly of geometric character and are easy to apply.

Two new approaches to function minimization which depart from the conventional quasi-Newton route were presented in Chapter 6. The approaches have already yielded a number of efficient numerical methods which are described elsewhere.

Looking at the monograph in perspective we feel that we have accomplished our goal of indicating interesting extensions of linear-quadratic control, matrix and optimization theory. We hope that this modest volume in the Bellman series will further stimulate research into extensions of the now well-established and well-known linear-quadratic theories.

Challenging areas for further research arise out of Chapters 3 and 4. In these chapters sufficient conditions for the existence of finite escape times for quadratic differential equations and for the non-negativity of constrained, non-linear functionals were presented. However, it is not known whether these conditions can be modified so as to be both necessary and sufficient. Only additional detailed research will resolve this question.

It should be possible to extend the results of Chapter 5 to non-autonomous systems though the situation is likely to be more complicated than in the autonomous case.

AUTHOR INDEX

Numbers in parentheses are reference numbers and indicate that an author's work is referred to, although his name is not cited in the text. Numbers in italics are the pages on which the complete references are listed.

SUBJECT INDEX